D1400113

The Quality Audit

The Quality Audit

A Management Evaluation Tool

Charles A. Mills P. Eng., C. Eng.

Sponsored by
American Society for Quality Control
Quality Audit Technical Committee

ASQC Quality Press, Milwaukee

Boston, Massachusetts Burr Ridge, Illinois
Dubuque, Iowa Madison, Wisconsin New York, New York
San Francisco, California St. Louis, Missouri

Library of Congress Cataloging-in-Publication Data

Mills, Charles A. (Charles Albert), date.
    The quality audit.

    Bibliography: p.
    Includes index.
    1. Quality control—Auditing. I. American Society
for Quality Control. II. Title.
TS156.M52 1989   658.5'62   88-32569
ISBN 0-07-042428-4

WISSER MEMORIAL LIBRARY

TS156
.M52
1989
copy1

McGraw-Hill

A Division of The McGraw·Hill Companies

Copyright © 1989 by McGraw-Hill, Inc. All rights reserved. Printed in
the United States of America. Except as permitted under the United
States Copyright Act of 1976, no part of this publication may be
reproduced or distributed in any form or by any means, or stored in a
data base or retrieval system, without the prior written permission of
the publisher.

11 12 13 14 BKMBKM 9 9 8

ISBN 0-07-042428-4

The editors for Quality Press were Tammy Griffin and Jeanine L. Lau.

The editors for this book were Robert Hauserman and Lester Strong,
and the production supervisor was Richard A. Ausburn. The book was
set in Century Schoolbook by the McGraw-Hill Publishing Company
Professional & Reference Division composition unit.

Information contained in this work has been obtained by McGraw-
Hill, Inc., from sources believed to be reliable. However, neither
McGraw-Hill nor its authors guarantess the accuracy or completeness of any information published herein and neither McGraw-Hill
nor its authors shall be responsible for any errors, omissions, or
damages arising out of use of this information. This work is
published with the understanding that McGraw-Hill and its authors
are supplying information but are not attempting to render engineering or other professional services. If such services are required,
the assistance of an appropriate professional should be sought.

For more information about other McGraw-Hill materials
call 1-800-2-MCGRAW in the United States. In other
countries, call your nearest McGraw-Hill office.

This book is dedicated to
my wife Peggy, to Jessie and Ruth,
and
to fellow members of the
Quality Audit Technical Committee,
American Society for Quality Control

Contents

Chapter 14. Reporting the Quality Audit   233

Chapter 15. Measuring and Improving the Effectiveness of the Quality Audit   267

Chapter 16. Conclusions   285

Preface

Most industries were introduced to the use of quality audits through the requirements of the U.S. Department of Defense Specification MIL-Q-9858 and similar military documents in other countries for:

1. Management review of quality programs
2. Precontract and periodic evaluation of the quality programs of suppliers

In some cases this introduction resulted from direct contracts with the military agencies and in others through the procurement chain starting with the prime contractors. Out of these basic requirements developed the more sophisticated internal and external quality audits of today.

I was introduced to the quality audit field in 1959 when I became manager of quality control at the Electronic Systems Division of Westinghouse Canada Inc. We were developing a quality program to match the requirements of the Canadian Department of National Defence. In this, I was aided by suggestions and comments from members of the quality control department of Autonetics, the designers of the radar equipment being built in Canada.

The value of external quality audits rapidly became apparent. After the initial precontract evaluation of suppliers, periodic audits were used to confirm the adherence of the suppliers to their stated quality programs. As our confidence developed in some suppliers, we treated their quality programs as extensions of our own and accepted their products for use with no further inspection, providing each shipment was accompanied by predefined technical data demonstrating compliance with the contract involved. Thus, the full value of the external quality audit could be demonstrated. A combination of source surveillance and incoming inspection was used with the remainder of the suppliers.

Initially the audit portion of the management review was carried out on an annual basis by selected supervisors and engineers from Westinghouse's quality department. The results were reported to upper management and any necessary corrective actions implemented. As the program developed, it became apparent that quality audits could also be used to evaluate the quality-oriented decisions being made by the inspection and test personnel of the quality control department itself. So decision sampling was born.

As I analyzed the product development system, it rapidly became apparent that the only way in which quality control could control product quality was through sorting operations at critical points in the product flow. This, of course, is counterproductive as a procedure and a costly misapplication of effort. In the mid-1960s I suggested to our upper management that product quality depended upon the line organizations, Design, Material Control, Production, etc., carrying out their activities right the first time. If this responsibility was recognized, the quality organization could act as a resource center to help each function develop its quality program and, using the Quality Audit, confirm that the quality decisions made in the line organizations were correct.

Consensus was reached on this responsibility and accountability in 1967, when it was decided that each line function would be responsible and accountable for the quality of its own work. Line personnel would be required to demonstrate that they had complied with the requirements of the product quality plan, drawings, specifications, and other technical data. All inspection and test functions were moved into the production activities, and a continuing form of quality audit introduced to monitor the various quality decisions and confirm their correctness. Shipments were authorized on the basis of the audit results. The responsibility for the initial external quality audit of a potential supplier remained with the quality department, but subsequent periodic audits of suppliers became the responsibility of the procurement department.

These methods of ensuring quality were used in regard to both military and high-technology commercial products by the Electronic Systems Division of Westinghouse Canada Inc. from 1967 until my retirement. Throughout all of this time, military approval of the company's quality program, to the highest levels demanded by NATO's quality standards, was maintained on a wide variety of products, and a high degree of customer satisfaction with the commercial products was attained as well. In all product lines, quality and reliability formed part of the marketing strategy.

The experimentation with and development of the various techniques discussed in this text involved patience and understanding on the part of various quality auditors, including C. J. Williams, P. Eng., and T. K. Rodgers, as well as the late Carl Northrop, P. Eng., and George Neil. The continued encouragement of management and others is also appreciated.

My participation in the activities of the American Society for Quality Control, the Institute of Quality Assurance, and the Canadian Standards Association also contributed significantly to the development of my thinking about quality audit techniques. As a founding member of the Quality Audit Technical Committee of ASQC, I found the discussions with fellow members most helpful.

Permission granted by the various standards writing organizations to use excerpts from their standards is very much appreciated. The copyrights for the quotations used remain with the organizations listed in Chapter 1, Appendix 1A, Section 7. Copies of the standards referenced in that appendix can be obtained directly from the issuing agencies or their agents. Responsibility for the interpretation of the excerpts used in this text rests with the author.

The preparation of the text owes much to the support, patience, and editorial assistance of my wife Peggy, and the patience of my granddaughters, Jessie and Ruth, who understood why so much time was being spent at the computer.

The staff at Quality Press was most helpful, particularly Jeanine Lau and Tammy Griffin, whose continual encouragement and patience is much appreciated.

One final note: Appreciating, as I do, the important and continuing contributions made by women to the field of quality control, every effort has been made to use gender-neutral language in this book. However, no generally accepted gender-neutral equivalents exist for the terms "foreman" and "workmanship," so they have been retained; in addition, non-gender-neutral terminology used in direct quotations from other sources has not been altered. In those instances, the words should be taken in a purely generic sense, intended to apply to women as well as to men.

A draft standard ISO 100011 on the Quality Audit, prepared by ISO Technical Committee TC 176 on Quality Assurance, is currently undergoing review and balloting by member countries. If approved, this document will undoubtedly be accepted by most member countries as their national standard for quality audits.

This standard's requirements address many of the headings included in this text. I find that it is not in conflict with any principles or techniques discussed in this books.

Charles A. Mills
February 1990

1

The Quality Audit

1.1 Introduction

The quality audit is a management tool used to evaluate, confirm, or verify activities related to quality. A properly conducted quality audit is a positive and constructive process. It helps prevent problems in the organization being audited through the identification of activities liable to create future problems. Problems generally arise through the inefficiency or inadequacy of the activity concerned.

The quality audit has negative connotations for some people. These feelings develop through the abuse of the audit process. This results from its use as a means of assigning blame or determining punitive actions.

1.2 Quality Audit—What Is It?

In determining "what the quality audit is," the obvious starting point is an accepted international or national standard that contains a glossary of terms or requirements defining the meaning of the words involved.

A review of a variety of English language quality standards and guides reveals many variations in the definitions and requirements for terms such as "audit" or "quality audit," as well as for other terms involving audit activities either wholly or in part. These other terms include "reviews" and "surveillance." In a number of cases, descriptive adjectives have been used to modify the general meaning. A summary of these is given in Appendix 1A.

International standard ISO 8402-1986, titled *Quality Vocabulary* (see Appendix 1A.3.1.1), defines the quality audit as follows:

> **Quality Audit**—A systematic and independent examination to determine whether quality activities and related results comply with planned

arrangements and whether these arrangements are implemented effectively and are suitable to achieve objectives.

This standard is issued by the International Organization for Standardization (ISO). It was prepared and approved by Technical Committee TC 176 on Quality Assurance. This committee contains a very broad representation of national standardization bodies (Appendix 1B). Its initial activities have all been associated with quality systems. Working groups are now active in other fields, including the development of an international standard on the quality audit. Such a standard will undoubtedly use the national standards of the United States and Canada as its foundation documents, as these are the only national standards on this topic. The work of this group of experts in the quality audit field could result in a new definition of the term "quality audit," more in line with audit standards and practices than the definition given above.

In my opinion, the present ISO definition does not adequately emphasize the fact that an audit examines and evaluates the acts and decisions of people involved with the quality system in its broadest sense.

Note: The term "quality system" is used in the comprehensive sense, i.e., covering the documentation and implementation of all activities having a bearing on the quality of the product, service, or process supplied by an organization.

An examination of the definitions and requirements discussed in Appendix 1A clearly shows the need for a more comprehensive definition of "quality audit" that could reduce or eliminate the many variations present. The definition given in the American standard ANSI/ASQC Q1-1986 and Canadian standard CAN-CSA-Q395-1981 (Appendix 1A.3.1.2) is more usable, but it is still not completely satisfactory.

As we have seen, the word "audit" is very common in the English language, having a myriad of meanings, each depending upon the application involved and in many cases the user. CAN-CSA-Q395-1981 defines the term as follows:

> **Audit**—A human evaluation process to determine the degree of adherence to prescribed norms (criteria, standards) and resulting in a judgement.

The audit is a long-established and well-respected activity in the accounting profession. It is used with a large number of modifiers and prefixes, including such terms as "system," "in-plant," "analytical," "data processing," etc.

Because of many similarities in the activities of the accounting and quality fields, quality professionals have adopted the same word "audit," complete with some of the same modifiers. In addition, new modifiers have been adopted to cover aspects peculiar to the quality profession. These modifiers are discussed in Chapter 2, "Types of Quality Audits."

The quality audit is a key factor in the management of the quality system of any organization, since it provides the data for evaluating and improving the effectiveness of that system. It is also the fundamental technique used for a "management review" or "management audit," frequently required by national and international procurement quality standards (see Appendix 1A.4). Management reviews and audits are internal management audit activities normally delegated to qualified internal quality auditors. The findings of these auditors are then reviewed with the management personnel concerned. Formal recognition of this delegation is found only in the Australian Standard AS2000-1978 (Appendix 1A.3.1.2).

Similarly, performance reviews (Appendix 1A.5) and some definitions of surveillance (Appendix 1A.6) have very strong audit connotations.

Because the term "quality audit" is so important, it is essential that its definition be clearly understandable and unambiguous to all concerned. To achieve these objectives, a definition has been developed from first principles.

The Oxford dictionary defines "audit" as a noun and verb, both applicable to our field:

1. *Noun.* An official examination of accounts with verification by reference to witnesses and vouchers.

2. *Verb.* To make an official systematic examination of accounts.

The first key phrase in the above definitions is "official systematic examination." This implies that the audit function must have an officially recognized position in the hierarchy of the organization or in relation to that organization. It also requires the audit to be "systematic," which in turn demands a planned, organized, and effective approach to carrying it out, i.e., a well-managed activity.

The second key phrase is "verification by references to witnesses and vouchers." This implies that the audit must involve itself with people and records. This combination of data sources is extremely pertinent in the quality field. It is this combination that separates the audit from an inspection, which is the verification of the product, service, or process being supplied. The word "verified" requires the findings to be based on confirmed facts and not hearsay evidence.

From the above, basic definitions of "quality audit" can be developed in the following forms:

1. *Noun.*   An official and systematic examination of the quality system with verification by witnesses and records.
2. *Verb.*   To make an official and systematic examination of the quality system with verification by witnesses and records.

These definitions rely on a common understanding of the word "quality," a word with a variety of shades and meanings. To ensure a common understanding, I will use the following definition given in ISO 8402-1986:

> **Quality**—The totality of features and characteristics of a product or service that bear on its ability to satisfy given needs.

In effect, this definition can be summarized as: putting the *right product* or *service* in the hands of the customer at the *right time* and at the *right price*. The phrase "right product or service" implies that the product or service, delivered to the customer, will satisfy the needs of that customer under the various conditions to be expected and for the desired operating life.

It has been accepted by most quality professionals that a quality product or service can only be achieved through deliberate and conscious acts and/or decisions by the people involved. All these acts and decisions must be targeted at achieving the desired quality. Quality does *not* result from happenstance or serendipity.

On the other hand, falldown in quality normally results from carelessness, oversight, misunderstanding, lack of knowledge, skills, or proper facilities, lack of direction, etc., in the performance of the particular task. Conscious or deliberate acts causing a falldown in quality constitute sabotage, normally a rare and isolated event.

Bringing these factors together, we can provide a more complete definition of the quality audit:

1. *Noun.*   An official, systematic examination and evaluation of the acts and/or decisions of people, taken to ensure that the applicable system, program, product, service, process, etc., meets all the required or desired characteristics, criteria, parameters, etc. This examination and evaluation is made by reference to witnesses and records. Therefore, it examines the intended and actual methodology, together with the results of that methodology. The evaluation

must cover the suitability, development, and implementation of each element concerned.

2. *Verb.*   To conduct a quality audit.

The noun definition very closely approximates that used in ANSI/ASQC A3-1978 (Appendix 1A.3.1.2). However, it has been expanded to more clearly define certain of the activities and their objectives. It also specifically requires the auditor to both examine and evaluate the activities concerned. This clarifies and strengthens the evaluation responsibilities placed on the auditor.

Numerous prefixes have been assigned to all the definitions of the quality audit. These describe the objectives of such audits. Typical prefixes include "management," "quality system," "quality program," "quality management," "product," "service," "process," etc. These are discussed in more detail in Chapter 2.

The examination and evaluation carried out during the quality audit will, in general, be used to determine one or more of the following:

1. The suitability of the documentation as it applies to the particular prefix, i.e., "system," "product," "service," "process," "management," etc.

2. The conformity or compliance of the operations to the established documentation.

3. The effectiveness of the quality system, i.e., of the documentation and its implementation.

Thus, the audit is concerned with both the methodology and the results of that methodology. An audit to evaluate conformity must include an assessment of the effectiveness of the methodology. These aspects are discussed in more detail in Chapter 2.

In summary, the quality audit evaluates the documentation and the resulting operations with respect to some predefined standard or specification. Its output is the report of its observations, plus, in some instances, requests for specific corrective actions.

1.3   Quality Audit—What It Is not

It can be categorically stated that the quality audit is *not* an alternative to an inspection operation. It can also be stated, unequivocally, that the quality audit *cannot* serve as a crutch to an ineffective inspection or quality program.

An inspection or test program is a decision-making process involving either an accept or reject decision, i.e., the product, service, or process does or does not conform to certain predefined criteria. The decisions are based on the comparison of predefined characteristics or parameters, forming the acceptance criteria, with those present in the item(s) under review. This comparison may involve visual observations, gauges, instruments, etc.

In most industrial situations, these decisions are made first by the operator carrying out the particular operation, and then verified by an independent inspector, i.e., an individual other than the one carrying out the actual operation being verified. This latter is the inspection function regardless of the title of the individual doing the verifying or that individual's reporting line in the organization.

The inspection function is normally associated with the material and manufacturing activities and their output. It is not normally associated with activities related to marketing, design, data processing, etc., nor with procedural activities, although possibly it should be as a number of activities in these areas are actually inspection functions, e.g., checking, verifying, or approving drawings and other documentation, proofreading, etc.

The inspection operation may also involve special inspection or test techniques carried out periodically to confirm that there has been no degradation of product, service, or process. Typical activities include periodic environmental tests, weld strength tests, x-rays, and other forms of periodic nondestructive and destructive evaluation tests, etc. If the tests are destructive or cause degradation, small sample sizes are frequently used, and since these are involved in determining the conformity of the product, service, or process to predefined requirements, they are inspection operations and not audit functions.

Inspection frequently uses sampling procedures to determine compliance or noncompliance with the requirements, albeit with the known Consumer's and Manufacturer's Risks associated with such procedures. Sampling does not make these kinds of inspections into audit activities.

The quality audit is a fact-finding process used to determine the suitability of, and/or the conformity to, the various quality program elements or requirements associated with the overall organization or company, elements of that organization, its product, service, or process output, etc. When evaluating inspection or test activities, it must be concerned with both the accept and reject decisions; however, it must not be involved with making these actual decisions.

The auditor may use inspection techniques as an evaluation tool. For example, in determining the effectiveness of an inspection procedure, the auditor may require measurements to validate accept and reject decisions. The review of the facts may result in an accept or reject decision. However, this decision will be with respect to the documentation and implementation of the quality system. It will not be based solely on the inspection results. The decision will be based on an evaluation of these and all other pertinent observations. This evaluation may result in further investigations or examinations in order to provide adequate evidence for the necessary decisions.

The quality audit should not be involved in carrying out any verification activities leading to the actual acceptance or rejection of the product or service. It should be involved with the evaluation of the process and controls covering the production and verification activities. In some cases, movement of goods may be dependent upon an approval of the decision-making ability of the inspectors concerned (see Section 10.2, "Decision Sampling").

1.4   Quality Audit—Why Do It?

Having defined what the quality audit is and what it is not, it is important to recognize the purpose—the "why"—of it.

The quality system of any organization is an integrated program of activities introduced by management, either of its own volition or as the result of a customer-imposed procurement quality standard. In either case, management must have some means of determining the effectiveness of the existing system and identifying areas requiring correction or improvement. The quality audit provides this means.

The quality audit is a management tool for determining the effectiveness of the quality system. The quality system may be that of management's own organization, a potential supplier, an active supplier, or an independent organization. The results of the audit provide an assessment of the adequacy of the existing program. They also provide a bench mark against which system improvements can be developed and evaluated.

Quality audits provide objective evidence to the management of the organization being audited and of the organization requesting the audit about the suitability of, conformity to requirements of, and effectiveness of the various elements of the quality system. The two management activities may or may not be synonomous, depending on the client/auditee relationship (see Section 1.4).

An internal quality audit, i.e., one conducted by employees of the organization being audited, should *not* be introduced just to satisfy a requirement of a procurement quality standard. Introduction on this basis will yield only marginal results. Management of the organization will not receive full value from the audit, as it will be an imposed extra rather than an integral part of the management system.

1.5   Quality Audit—Who Are Involved

1.5.1   General

The quality audit involves three functional parties that may be related in a number of ways. Functionally these are defined by CAN-CSA-Q395-1981 as:

> **Auditor**—A person qualified to plan and conduct audits in accordance with this Standard; and in this Standard may be used in reference to one or more auditors.

> **Client**—The organization that requested the auditing organization to conduct the audit.

> **Auditee**—The organization to be audited (the auditee may be a division of the client or an entirely separate organization).

In examining each of these terms in more detail, it becomes readily apparent that the interrelationship can be much more complex than these somewhat simplified definitions imply.

1.5.2   Client

The client is the organization or individual requesting that the quality audit be carried out. In making this request, the client must clearly define the purpose of the audit and the standard of performance against which the auditor's findings will be compared. These reference documents can include quality system standards, product standards, process standards, good manufacturing practices (GMP), etc. These are discussed in more detail in Chapter 2.

A list of typical clients includes the following:

1. *Potential customer.* A potential customer may wish to evaluate the ability of an interested supplier to provide products or services that will meet the needs of that customer. In this case, the evaluation would be against one or more of the following—procurement quality standard, product, service, or process standard, standard practices, etc. This type of precontract evaluation is a requirement of many commercial and military procurement quality standards and operating practices.

2. *Customer.* A customer may request a quality audit of a supplier as part of the continuing evaluation of suppliers to detect any degradation of the quality system, product, service, process, etc. Or it may be used to detect the cause of some degradation which has arisen. This continuing evaluation is a requirement of many commercial and military procurement quality standards.

3. *Organization management—quality system evaluation.* The management of an organization may request an outside organization or specialists to evaluate and report on the suitability and effectiveness of its quality program as it applies overall or to particular products, services, or processes, i.e., an external quality audit. This type of request may arise from a recognition of the need for an improved quality program, from a decision to use quality as a marketing strategy, from a desire to enter a particular marketplace having specific quality requirements, etc.

4. *Organization management—internal quality audit.* The management of an organization may request its own internal quality department to carry out quality audits of the activities within the organization. This audit would be to determine the degree of compliance by the operation with its defined operating procedures and to assess the effectiveness of the quality program. This type of audit is required by some procurement quality standards.

5. *Organization management—quality system registration.* The management of an organization may request a registration or certification agency to conduct an audit of its quality system against some agreed upon standard, with a view to receiving approval and being included on a published list of approved suppliers. Several countries have such programs applying to quality systems and/or products, services, processes, and so forth—e.g., for commercial concerns in Great Britain, Canada, Eire, and New Zealand and for military purposes in a number of the NATO countries.

6. *Regulatory agencies.* A regulatory or military agency will frequently require a potential supplier or, in particular fields, a user to obtain approval of its quality system, operating system, product, service, or process controls, etc., prior to receiving authorization to commence manufacturing or providing a particular product, service, or process. Typical users of this technique include military or defense, nuclear, food, and drug agencies.

1.5.3 Auditor

All quality auditors, organizational or individual, fall into one of two major categories—external or internal. These can be defined as:

1. *External quality auditors.* External quality auditors are not members of the organization being audited. They may be:
    a. A third-party organization or individuals hired by the client to conduct the audit on its behalf.

b. A third-party organization or individuals hired by an approval agency to conduct the audit, initiated at the request of the client.

c. Second-party auditors, in the employment of the customer, potential customer, or other independent organization requesting the audit of the auditee.

d. Second-party auditors, employed by an approval agency, carrying out an audit to determine the ability of the auditee to provide the desired quality system, product, service, or process.

e. Auditors, employed by a corporate headquarters, carrying out an audit to determine whether a division or other element of that corporation complies with corporate policies and desires.

2. *Internal quality auditors.* Internal quality auditors are employees of the organization being audited.

The implications of internal and external quality audits are broader than just the individuals carrying them out. However, to a large extent the other implications tend to result from the interrelationships between auditors and auditees. These are discussed in more detail in Chapter 2.

1.5.4 Auditee

The auditee is the organization being audited, regardless of its relationship with either the client or the auditor. In the vernacular, the auditee is "on the receiving end of the activity."

The auditee may be a complete organization, a major element or segment of an organization (e.g., a division), or a minor segment of an organization (e.g., a particular product, service, or process, a cost center, or a specialized activity, etc.).

1.6 Quality Audit—Where Is It Conducted?

On the surface, this question would appear to be superfluous as undoubtedly most of the audit must be conducted on the premises of the auditee, since that is where the evidence exists. However, this may not be the most effective, or even practical, location for carrying out all aspects of the quality audit.

In determining whether a quality system is satisfactory or not, there are two phases to the activity. The first is to determine the suitability of the documentation on the system with respect to the reference standard. The second is to determine the conformity of the various activities to the documentation and the effectiveness of that implementation.

Since the presence of auditors in a work area will cause some disruption, it is frequently more effective to carry out the suitability quality audit off the auditee's premises. The most cost-effective location will frequently be the base or home office of the auditors.

In evaluating the suitability of a product, service, or process, specialized test equipment is frequently required that may not be available on the auditee's premises. In this case, the auditor will frequently use facilities owned or leased by the client.

It may also be necessary to gather evidence on the effectiveness of the quality system from customers or users of the product, service, or process. The location for these reviews will depend upon the nature and whereabouts of the evidence in relation to the parties involved.

Evidence on material control may require visits to suppliers or vendors to determine the adequacy of the quality system in dealing with these parties.

1.7 Quality Audit—When Should It Be Performed?

The quality audit may be a single occurrence or a repetitive activity, depending on the purpose and the results of both the audit and the quality system, product, service, or process concerned.

Virtually all quality audits are repetitive to some extent, for the following reasons:

1. Most external quality audits require some form of followup audit. On a short-term basis, the followup involves reviewing the implementation of any requested corrective action to assess its suitability and effectiveness. Longer-term followup involves periodic audits to confirm there has been no degradation in the quality system from the level originally approved.

2. Virtually all internal quality audits are repetitive, with the periodicity depending on the nature of the activity being audited, the modus operandi of the organization, and the desires of the client. The periodicity may vary from daily up to once a year, depending largely on the criticality of the product and the quality control system. Certain of the procurement quality standards state a maximum period between complete internal audits. Similarly, some product, service, and process standards define the periodicity of audits and/or special verification activities.

A single-occurrence quality audit will normally occur only where the quality system is found to be unsatisfactory and the auditee has shown no intention of correcting the shortcomings detected.

Appendix 1A

Definitions of Quality Audit and Associated Terminology from International and National Standards

1A.1   Introduction

A review was made of the major English language standards (see Appendix 1A.7 below) associated with the fields of quality assurance and the quality audit. These included the national standards of the United States, Canada, the United Kingdom, and Australia and the international standards issued by the International Organization for Standardization and the North Atlantic Treaty Organization (NATO). Certain of the national standards have been harmonized with the ISO standards or issued under a dual-numbering system, as noted in Appendix 1A.7.

Appropriate definitions and requirements have been extracted. These definitions have been divided under the following headings:

1. Audits—Section 1A.2

2. Quality audit—Section 1A.3

3. Reviews—Section 1A.4

4. Performance reporting—Section 1A.5

5. Surveillance—Section 1A.6

1A.2   Audits

1A.2.1   Definitions

The following definition of the term "audit" was taken from the Canadian national standard CAN-CSA-Q395-1981, *Quality Audits:*

> **Audit**—A human evaluation process to determine the degree of adherence to prescribed norms (criteria, standards) and resulting in a judgement.

1A.2.2   Requirements

In addition, certain quality system standards lay down requirements or guidelines for the audit with no formal definition, including:

1. American national standard ANSI/ASME NQA-1-1986, *Quality Assurance Program Requirements for Nuclear Facilities:*

Audits—Planned and scheduled audits shall be performed to verify compliance with all aspects of the quality assurance program and to determine its effectiveness. These audits shall be performed in accordance with written procedures or checklists by personnel who do not have direct responsibility for performing the activities being audited. Audit results shall be documented and reported to and reviewed by responsible management. Follow-up action shall be taken where indicated.

2. Australian national standards AS 1821 to AS 1823-1985, *Supplier Quality Systems:*

   Review and Audit—The quality system shall be:

   a) periodically and systematically reviewed and audited by the supplier to ensure continued effectiveness;

   b) subject to periodic review and audit by the quality assurance representative. Resulting from such review and audit, the system or any of its elements may be disapproved. Reason(s) for this shall be stated in writing and modifications to correct deficiencies may be required.

1A.3 Quality Audit

1A.3.1 Basic quality audit

The following definitions were noted for the term "quality audit":

1. International standard ISO 8402-1986, *Quality Vocabulary:*

   Quality Audit—A systematic and independent examination to determine whether quality activities and related results comply with planned arrangements and whether these arrangements are implemented effectively and are suitable to achieve objectives.

2. National standard of Canada CAN-CSA-Q395-1981, *Quality Audits* and American national standard ANSI/ASQC-Q1-1986, *Generic Guidelines for Auditing Quality Systems:*

   Quality Audit—A systematic examination of the acts and decisions of people with respect to quality in order to independently verify or evaluate and report the degree of compliance to the operational requirements of the quality program, or the specification or contract requirements of the product or service.

3. American national standard ANSI/ASQC-A3-1987, *Quality Systems Terminology:*

   Quality Audit—A systematic and independent examination and evaluation to determine whether quality activities and results comply with planned arrangements and whether these arrangements are implemented effectively and are suitable to achieve objectives.

4. National standard of Canada CAN-CSA-Z299.1-1985 (originally numbered CAN3-Z299.1-1985), *Quality Assurance Program Category 1:*

> **Quality Audit**—a documented activity aimed at verifying by examination and evaluation that the applicable elements of the quality assurance program have been established, documented and implemented effectively in accordance with specified requirements.

1A.3.2 Modified definitions

In addition to the above basic definitions, some standards provide definitions of "quality audit" coupled with a modifier describing the particular application.

1. American national standard ANSI/ASQC-Q1-1986, *Generic Guidelines for Auditing Quality System:*

> **Quality System Audit**—is a documented activity performed to verify, by examination and evaluations of objective evidence that applicable elements of the quality system are appropriate and have been developed, documented and effectively implemented in accordance and in conjunction with specified requirements.

Note: "Quality system" is defined in this standard as:

> ...the documented plans, organizational structure and activities that are implemented to control the conformance of a product or service to specified requirements and to provide evidence of such conformance.

2. National standard of Canada CAN-CSA-Q395-1981, *Quality Audits:*

> **Quality Program Audit**—means the documented activity performed to verify, by examination and evaluation of objective evidence, that applicable elements of the quality program have been developed, documented, and effectively implemented in accordance with specified requirements.

Note: "Quality program" is defined in this standard as:

> ...the documented plans, organizational structure, and activities that are implemented to control the conformance of a product or service to specified requirements and to provide evidence of such conformance.

Note: The definitions in the above ASQC and CSA standards are virtually synonymous except for the interchange of the terms "quality system" and "quality program." This difference is one of the reasons for ISO 8402-1986 recognizing the two terms as themselves being synonomous.

Other definitions from CAN-CSA-Q395-1981 include:

Process Quality Audit—an analysis of a process and appraisal of completeness and correctness of the conditions with respect to some standard.

Product Quality Audit—A quantitative assessment of conformance to required product characteristics

3. International standard ISO 9001-1987, *Quality Systems—Model for Quality Assurance in Design/Development, Production, Installation and Servicing:*

Internal Quality Audits—The supplier shall carry out a comprehensive system of planned and documented internal quality audits to verify whether quality activities comply with planned arrangements and to determine the effectiveness of the quality system.

Audits shall be scheduled on the basis of the status and importance of the activity.

The results of the audits shall be documented and brought to the attention of the personnel having responsibility in the area audited. The management personnel responsible for the area shall take timely corrective action on the deficiencies found by the audit team.

4. International standard ISO 9002-1987, *Quality Systems—Model for Quality Assurance in Production and Installation:*

Internal Quality Audits—The supplier shall carry out internal quality audits to verify whether quality activities comply with planned arrangements and to determine the effectiveness of the quality system.

Audits shall be scheduled on the basis of the status and importance of the activity.

The audits and follow-up action shall be carried out in accordance with documented procedures.

The results of the audits shall be documented and brought to the attention of the personnel having responsibility in the area audited. The management personnel responsible for the area shall take timely corrective action on the deficiencies found by the audit.

5. International standard ISO 9004-1987, *Quality Management and Quality System Elements—Guidelines:*

Auditing the Quality System—General—All elements and components pertaining to a quality system should be internally audited and evaluated on a regular basis. Audits should be carried out in order to determine whether various elements within a quality management system are effective in achieving stated quality objectives. For this purpose, an appropriate audit plan should be formulated and established by company management.

Note: This paragraph is followed by more detailed recommendations for the audit plan, carrying out the plan, reporting, and following up of audit findings.

6. Australian standard AS 2000-1978, *Guide to AS 1821–1823 Supplier's Quality Control Systems:*

> **Management Audit**—may be defined as a planned, purposeful and comprehensive examination of management objectives, assignments of duties, delegation of responsibilities and methods of operation. Such audits are conducted by, or on behalf of, management to check that these objectives, delegations and methods are achieving the required results, to reveal defects or irregularities in any of the elements examined and to indicate possible improvements. Audits serve as a check on the abilities of management at all levels. They are designed to uncover potential danger spots and to eliminate waste or unnecessary loss. The requirements for a management audit should not be interpreted to mean top management must personally conduct the audit. This task may, and undoubtedly will, be delegated.

7. American national standard ANSI/ASQC Z1.15-1979, *Generic Guidelines for Quality Systems:*

> **Quality Systems Audits**—To provide assurance, a periodic audit of the quality system should be made by an organizational element independent of the unit being audited or by a qualified third party. It may include as appropriate:
>
> 1. **Management Audits** to determine how well quality policy and objectives are being met.
> 2. **Systems Audits,** including manufacturing process audits, to determine how well quality planning has been implemented and to identify areas where changes would be beneficial to product quality and costs.
> 3. **Product Audits** to determine how well products, after manufacture, inspection and test meet quality requirements.

8. American national standard ANSI/ASQC A3-1987, *Quality System Terminology:*

> **Quality System Audit** (Quality Program Audit)—A documented activity performed to verify, by examination and evaluation of objective evidence, that applicable elements of the quality system are suitable and have been developed, documented and effectively implemented in accordance with specified requirements.

Note: "Quality system" is defined and used in ANSI/ASQC A3-1987 in the same manner as in ANSI/ASQC Q1-1986 (Reference 3.2.1):

Product Quality Audit—A quantitative assessment of the conformance to the required product characteristics.

Process Quality Audit—An analysis of elements of a process and appraisal of completeness, correctness of conditions, and probable effectiveness.

1A.4   Reviews

Some quality standards researched contain either definitions or requirements for a "review," either by title or wording of the text. These are in addition to, or in lieu of, quality audit requirements in some form. But in all cases, the actions are synonomous with those of a quality audit. Most documents require the review to be undertaken by management, with only the Australian standards AS 1821 to AS 1823-1985 and AS 2000-1978 recognizing that the audit will undoubtedly be delegated. However, even if not delegated, quality audit principles and techniques should apply.

1A.4.1   Definitions

The following was taken from International standard ISO 8402-1986, *Quality Vocabulary:*

> **Quality System Review**—A formal evaluation by top management of the status and adequacy of the quality system in relation to quality policy and new objectives resulting from changing circumstances.

1A.4.2   Requirements

The following requirements regarding reviews are taken from the standards listed below:

1. International standards ISO 9001-1987, *Quality Systems—Model for Quality Assurance in Design/Development, Production, Installation and Servicing,* ISO 9002-1987, *Quality Systems—Model for Quality Assurance in Production and Installation,* and ISO 9003-1987, *Quality Systems—Model for Quality Assurance in Final Inspection and Test:*

   > **Management Review**—The quality system adopted to satisfy the requirements of this International Standard shall be reviewed at appropriate intervals by the supplier's management to ensure its continuing suitability and effectiveness. Records of such reviews shall be maintained.

Note: In addition, ISO 9001-1987 and ISO 9002-1987 include the following note as part of the above discussions of reviews:

> **Note:** Management reviews normally include assessment of the results of internal quality audits, but are carried out by, or on behalf of the supplier's management, viz. management personnel having direct responsibility for the system.

2. International standard ISO 9004-1987, *Quality Management and Quality System Elements—Guidelines:*

> **Review and Evaluation of the Quality System**—Provision should be made by company management for independent reviews and evaluation of the quality system. Such reviews should be carried out by appropriate members of company management or by competent independent personnel as decided by company management.
>
> Reviews should consist of well structured and comprehensive evaluations which include
>
> a) findings of audits centred on various elements of the quality system
>
> b) the overall effectiveness of the quality management system in achieving stated objectives
>
> c) consideration for up-dating the quality management system in relation to changes brought about by new technologies, quality concepts, market strategies and social or environmental conditions
>
> Findings, conclusions and recommendations should be submitted in documentary form for necessary action by company management.

3. NATO standard AQAP-1, 3d ed., *NATO Requirements for an Industrial Quality Control System:*

> **Quality Control System Review**—Contractor management shall conduct periodic and systematic review to demonstrate the effectiveness of the system.

4. NATO standard AQAP-4, 3d ed., *NATO Requirements for an Industrial Inspection System:*

> **Review and Evaluation**—The inspection system established in accordance with the provisions of this document shall be periodically and systematically reviewed by the contractor to ensure its effectiveness and is subject to evaluation by the Quality Assurance Representative who may disapprove the system or any of its elements.

5. U.S. military specification MIL-Q-9858A-1963, *Quality Program Requirements:*

> **a) Organization**—Management regularly shall review the status and adequacy of the quality program.
> **b) Control of Purchases**—The effectiveness and integrity of the control

of quality by his suppliers shall be assessed and reviewed by the contractor at intervals consistent with the complexity and quality of product.

Note: U.S. Department of Defense handbook H50, *Evaluation of a Contractor's Quality Program,* gives the interpretation and methods of conducting assessments to and under MIL-Q-9858A.

6. British national standard BS 5750-1987, parts 1, 2, and 3, have been issued as dual-numbered standards with ISO 9001, 9002, and 9003, respectively, and thus use the ISO definitions and requirements.

7. National standard of Canada CAN-CSA-Z299.1-1985, *Quality Assurance Program—Category 1:*

   Management Review—Provide for the regular review by senior management of the status and adequacy of the quality assurance program to ensure its continuing suitability and effectiveness in meeting the requirements of this standard.

8. American national standard ANSI/ASME NQA-1-1986, *Quality Assurance Program Requirements for Nuclear Facilities:*

   Management of those organizations implementing the quality assurance program or portions thereof, shall regularly assess the adequacy of that part of the program for which they are responsible and shall assure its effective implementation.

9. Australian standards AS 1821 to AS 1823-1985, *Supplier Quality Systems,* have a review requirement (Appendix 1A.2.2.2).

1A.5 Performance Reporting

One standard describes the responsibility of reporting on the performance of the quality system as it compares to stated quality requirements or objectives. That requirement for "performance reporting" is given in American national standard ANSI/ASQC Z1.15-1979, *Generic Guidelines for Quality Systems:*

Quality Performance Reporting—Responsibility for reporting to higher management performance against stated quality objectives should rest with functions independent of those responsible for the attainment of those objectives. Procedures for documentation and record retention should be established.

1A.6 Surveillance

The following international and national standards provide either definitions of or requirements for "surveillance." In each case, activities are required that are synonomous with audit activities.

1. International standard ISO 8402-1986, *Quality Vocabulary:*

   Quality Surveillance—The continuing evaluation of the status of procedures, methods, conditions, products, processes and services and analysis of records to assure quality requirements will be met.

2. National standards of Canada CAN-CSA-Z299.1 through CAN-CSA-Z299.4, 1985 editions, titled, respectively, *Quality Assurance Program Requirements*, Categories 1 through 4:

   Surveillance—The continuing evaluation, analysis and verification of a supplier's records, methods, procedures, products and services, to assure that requirements are met.

3. American national standard ANSI/ASQC A3-1987, *Quality System Terminology:*

   Surveillance—Monitoring or observation to verify whether an item or activity conforms to specified requirements.

1A.7   Standards Reviewed

The extracts quoted in this appendix were taken from the designated editions of the standards listed below. Each quotation is reproduced with the kind permission of the issuing authority.

In addition to the standards quoted, certain other U.S., Canadian, and British standards were reviewed. These documents contained definitions and requirements identical to the ISO standards quoted. These are identified in the standards listed under the different standards writing organizations.

Each of the referenced standards or guides are copyrighted by the issuing authority, and hence the copyright for each quotation is held by that authority. All quotations have been extracted from the issue or revision of the standard or guide listed below. These documents are subject to amendment, as deemed necessary, by the issuing authority.

1A.7.1   American national standards—
American National Standards Institute/
American Society for Quality Control

Extracts have been made, with permission, from the following American national standards published by The American Society for Quality Control, 310 West Wisconsin Avenue, Milwaukee, WI 53203:

1. ANSI/ASQC Q1-1986, *Generic Guidelines for Auditing Quality Systems*
2. ANSI/ASQC A3-1978, *Quality Terminology*
3. ANSI/ASQC Z1.15-1979, *Generic Guidelines for Quality Systems*

In addition the following standards were reviewed for compatibility with the corresponding ISO standard:

1. ANSI/ASQC Q91-1987, *Quality Systems—Model for Quality Assurance in Design/Development, Production, Installation and Servicing;* technically equivalent to ISO 9001-1987
2. ANSI/ASQC Q92-1987, *Quality Systems—Model for Quality Assurance in Production and Servicing;* technically equivalent to ISO 9002-1987
3. ANSI/ASQC Q93-1987, *Quality System—Model for Quality Assurance in Final Inspection and Test;* technically equivalent to ISO 9003-1987
4. ANSI/ASQC Q94-1987, *Quality Management and Quality Systems Elements—Guideline;* technically equivalent to ISO 9004-1987

The copyright for the above standards is held by the American Society for Quality Control, Inc., Milwaukee, WI.

1A.7.2 American national standard— American National Standards Institute/ American Society of Mechanical Engineers

Extracts have been selected from the following American national standard published by The American Society of Mechanical Engineers, United Engineering Center, 345 East 47th Street, New York, NY 10017:

1. ANSI/ASME NQA1-1986, *Quality Assurance Program Requirements for Nuclear Facilities*

1A.7.3 Australian national standards— The Standards Association of Australia

Extracts have been selected from the following Australian national standards published by The Standards Association of Australia, Standards House, 80 Arthur Street, North Sydney, New South Wales, Australia:

1. AS 1821 through AS 1823-1985, *Suppliers Quality Systems*
2. AS 2000-1978, *Guide to AS 1821–1823 Suppliers Quality Control Systems*

1A.7.4   British national standards—
British Standards Institute

Reference has been made to the following British national standards published by The British Standards Institute, Linford Wood, Milton Keyes, MK14 6LE, United Kingdom. These standards are issued as dual-numbered documents carrying both BSI and ISO designations.

1. BS 5750-1987, Part 1, *Quality Systems—Model for Quality Assurance in Design/Development, Production, Installation and Servicing,* dual-numbered as ISO 9001-1987

2. BS 5750-1987, Part 2, *Quality System—Model for Quality Assurance in Production and Installation,* dual-numbered as ISO 9002-1987

3. BS 5750-1987, Part 3, *Quality System—Model for Quality Assurance in Final Inspection and Test,* dual-numbered as ISO 9003-1987

4. BS 5750-1987, Part 0, Section 2, *Quality Management and Quality System Elements—Guidelines,* dual-numbered as ISO 9004-1987

5. BS 4778-1987, Part 1, *International Terms,* dual-numbered as ISO 8402-1986

1A.7.5   Canadian national standards—
Canadian Standards Association

Extracts have been selected from the following Canadian national standards published by the Canadian Standards Association, 178 Rexdale Boulevard, Rexdale, Ontario, Canada M9W 1R3. The quotations listed are reproduced with the permission of the Canadian Standards Association from CSA Standards Q395-1981 and Z299.1 through Z299.4, which are copyrighted by CSA and may be subject to amendment or revision.

1. CAN-CSA-Q395-1981, *Quality Audits*

2. CAN-CSA-Z299.1 through CAN-CSA-Z299.4, *Quality Assurance Program Requirements,* Categories 1 through 4

In addition reference has been made to the following CSA standards incorporating the referenced ISO documents:

1. CAN-CSA-Q640-1987, *Quality Vocabulary,* incorporating ISO 8402–1986

2. CAN-CSA Q420-1987, *Quality Management Systems,* incorporating ISO 9004-1987

1A.7.6   International standards—
The International Organization for
Standardization

Extracts have been selected from the following international standards published by The International Organization for Standardization, Geneva, Switzerland, also available through the member national standards organization:

1. ISO 8402-1986, *Quality Vocabulary*
2. ISO 9000-1987, *Quality Management and Quality Assurance Standards—Guidelines for Selection and Use*
3. ISO 9001-1987, *Quality Systems—Model for Quality Assurance in Design/Development, Production, Installation and Servicing*
4. ISO 9002-1987, *Quality Systems—Model for Quality Assurance in Production and Installation*
5. ISO 9003-1987, *Quality Systems—Model for Quality Assurance in Final Inspection and Test*
6. ISO 9004-1987, *Quality Management and Quality System Elements—Guidelines*

1A.7.7   International military standards—
The North Atlantic Treaty Organization

Extracts have been selected from the following international military standards published by The North Atlantic Treaty Organization (NATO), and available through member national defense organizations:

1. AQAP-1, 3d ed., *NATO Requirements for an Industrial Quality Control System*
2. AQAP-4, 3d ed., *NATO Requirements for an Industrial Inspection System*

1A.7.8   American military standards—
U.S. Department of Defense

Extracts have been selected from the following United States defense documents published by The Office of the Assistant Secretary of Defense (Installations and Logistics), Washington, DC, 20301:

1. MIL-Q-9858A-1963, *Quality Program Requirements*
2. Handbook H50, *Evaluation of a Contractor's Quality Program*

Appendix 1B

International Organization for Standardization (ISO):

Membership in Technical Committee

ISO/TC 176 on Quality Assurance

The International Organization for Standardization (ISO) Technical Committee ISO/TC 176 is responsible for developing generic standards and guides in the field of quality assurance. This committee consists of representatives from over 50 national standards bodies that have expressed active interest in the development of these documents. Some are participating (P) members; others are observing (O) members.

Participating members (P members), 29 countries				
Argentina	Finland	Italy	Portugal	Switzerland
Australia	France	Jamaica	South Africa,	Trinidad and Tobago
Austria	Germany, F.R.	Japan	Republic of	United Kingdom
Belgium	India	Netherlands	Spain	USA
Brazil	Iran	Norway	Sri Lanka	USSR
Canada	Ireland	Peru	Sweden	Yugoslavia

Observing members (O members), 29 countries				
Chile	Denmark	Israel	New Zealand	Tanzania
China	Ghana	Ivory Coast	Philippines	Thailand
Colombia	Greece	Kenya	Romania	Tunisia
Cuba	Hungary	Malaysia	Saudi Arabia	Turkey
Cyprus	Iceland	Mauritius	Singapore	Venezuela
Czechoslovakia	Indonesia	Mexico	Syria	

2

Types of Quality Audits

2.1 Introduction

Quality audits can be subdivided into four categories or classifications based on:

1. The purpose of the audit—"Why?"
2. The object of the audit—"What?"
3. The nature of the audit—"Who?"
4. The method of the audit—"How?"

Each of these classifications subdivides into subsidiary categories, discussed in the following paragraphs. In practice, virtually all permutations and combinations of these categories can be found.

2.2 Quality System vs. Quality Program

There is considerable controversy and confusion over the meanings of the terms "quality system" and "quality program" and their interrelationship.

Both terms have been used to describe the documentation used to correlate the various activities that implement the quality policy of an organization. Both have also been used to refer to the documentation plus all the activities resulting from that documentation. These conflicting views are found when comparing the national standards of different countries. There does appear to be consistent usage within each standards writing organization. There also appears to be the need to differentiate between the two activities, as virtually all national standards reviewed use two terms to differentiate between the collective plans, documents, activities, etc., and the documentation on its own.

At the international level, ISO TC 176 has decided to treat the terms as synonomous, thereby neglecting the need for two terms to differentiate between the documentation on its own and the documentation plus the resulting operations.

However, despite this international approach, I will treat these terms as they are defined in ANSI/ASQC A3-1978 and use either "quality program" or "quality documentation" to cover the documentation element alone and "quality system" to cover the documentation plus all of the activities resulting from that documentation.

2.3   Purpose—"Why"—of the Quality Audit

Quality audits are carried out to determine either or both of the following:

1. Suitability of the quality program (documentation) with respect to a predetermined reference standard

2. Conformity of the operations within the quality system to the documented quality program

Thus the suitability quality audit validates the documentation, i.e., the planning aspects, of the quality system against the predetermined reference standard, while the conformity quality audit validates the activities of the quality system, i.e., the implementation, measurement, corrective action, and subsequent followup, against the documentation forming the quality program.

These types of audits may be more formally defined as:

1. *Suitability quality audit.* An audit or in-depth evaluation and comparison of the quality program (documentation) for the organization, specific elements of that organization, product, process, service, etc., against the reference standard, predetermined by the client

2. *Conformity quality audit.* An audit or in-depth evaluation and comparison of the activities within the quality system against the predefined quality program, i.e., against quality policies and procedures

2.4   Object—"What"—of the Quality Audit

The quality audit is used to determine the suitability and effectiveness of the quality system as it is applied to the whole or elements of an organization, product, process, service, etc. Thus we are dealing with the management of these activities in terms of the traditional management cycle:

The quality system involves all four phases of management, whereas the quality program is involved with the planning phase.

The object of the quality audit will normally be shown by a modifier preceding the phrase "quality audit." These modifiers can cover the full gamut of activities covered by a quality program, including system, management, product, process, service, etc. These can lead to further definitions with respect to the quality audit.

2.4.1 Quality program audit

The quality program includes the policies, procedures, operating instructions, etc., necessary to define the various responsibilities, accountabilities, and actions necessary to achieve the desired quality level. This level may be based on:

1. The desires of management

2. The needs of the intended marketplace

3. National or international standards for quality assurance

4. Procurement quality standards from major customers or potential customers

5. Good manufacturing practices as laid down by certain regulatory agencies

6. Specific requirements in the applicable product, service, or process specification or standard

The client selects the reference standard based on the purpose of the quality system. The reference standard may be any of the above mentioned documents or one suggested by an outside consultant as applicable to the important facets of a particular quality program.

The quality program, in its various forms, provides the policies, guidelines, and instructions for the various activities in the quality system and the reference standard for comparing and evaluating these activities. The formal documentation of the quality system, required by many international and national quality system/program standards, ensures that all concerned are working according to a common set of ground rules, whether those tasks involve carrying out a particular operation or activity, verifying the output of operations, or auditing the system's activities. There are a wide variety of names applied to this documentation, as well as a myriad of formats. The names and formats should be chosen to provide the most effective communication link between the parties concerned. The names or formats may, or may not, be those given in the reference standard. However, they must satisfy the needs of the reference standard. Regardless of the names or formats used, the importance in encouraging better discipline is the same.

The quality program audit is carried out to determine the suitability of the documents comprising the program. Therefore, it consists of a comparison between these documents with the requirements of the reference standard. Each requirement clause must be addressed by the documentation and must be reviewed by the auditor.

Thus we have the following definition:

> *Quality program audit*—a suitability audit or in-depth comparison and evaluation of the documentation comprising the quality program, for the organization or elements thereof, product, process, service, etc., against the agreed-on predetermined reference standard

2.4.2 Quality system audit

The quality system comprises the quality program plus all the activities and operations required to implement it effectively. Thus it covers the planning, implementation, measurement, and correction phases of the traditional management cycle:

This management cycle applies whether one is considering the overall organization or a particular element, product, process, or service within that organization.

As applied to Quality Systems, this cycle is actually a repetitive function operating as a converging spiral as new or more innovative techniques are developed for improving the system and the results of that system. The "constant improvement" philosophy applies as much to the quality system as to the output of the organization.

The initial cycle as shown above represents the functions necessary to introduce the initial elements of the quality system. Each correction phase results in a complete management cycle, i.e., the corrective action must be planned and implemented, its effectivity measured, and then it must be followed by further corrective action to further improve the operations.

As each correction cycle is repeated, the corrections required should be smaller and smaller and fewer and fewer until the desired quality level is achieved. From the point of view of product, zero defects is the desired quality level. In some cases, because of the sheer weight of numbers, this may only be achieved after much experimentation and improvement. However, as soon as possible departures from requirements should be measured in n parts per million (ppm), where n is normally expressed in terms of units or tens. Thus, process control techniques become significant tools in showing where a process is and where it is going. From the systems point of view, many decision-making points involve numerous decisions affecting quality, i.e., involve making valid quality decisions. Hence many of the control techniques used in the process control are equally applicable to improving the ability to make valid quality decisions.

Quality system audits can cover one or all of the following typical categories:

1. Performing a suitability quality audit on the quality program with respect to some predetermined quality system requirement; then, if the documentation is satisfactory, performing a conformity quality audit to determine the effectiveness of the resulting activities

2. Performing a suitability audit on the changes made to an approved quality program to confirm that it is still acceptable; if it is still suitable, performing a conformity quality audit to evaluate the changes and determine continuing compliance with the approved quality program

3. Performing repetitive or periodic conformity quality audits on the implementation of an approved quality program to determine continuing compliance and effectiveness

The choice of scenario will depend on the relationships between client, auditor, and auditee.

Thus we have the following definition:

> *System quality audit*—The system quality audit is an audit or in-depth examination of the quality system to determine the effectiveness and compliance of the system with the predetermined reference standard. An audit of this nature can include a suitability quality audit in addition to the implied conformity quality audit. The system quality audit examines and evaluates the quality system as it applies to an overall organization

or to a particular element, product, process, service, etc., within that organization.

2.4.3 Management quality audit

The term "management quality audit" is frequently used as an alternative to "system quality audit," in all its variations. However, more correctly, it should be restricted to the management aspects of the quality system as applied to an overall organization. This type of audit reviews and evaluates the responsibilities, accountabilities, actions, interactions, etc., of the management team with respect to all the activities contributing to the output quality of the organization. This type of audit applies equally to product- and service-oriented organizations.

The actions and interactions of the various functions within the organization will depend on the complexity of the quality system. In turn, the complexity of the system will depend on a number of factors, including:

1. The complexity of the applicable procurement quality standard
2. The criticality of the application of the item concerned in terms of operation, safety, etc.
3. The requirements of the intended marketplace
4. The maturity of the product, service, or process
5. The requirements of the particular product, process, or service standard or specification

2.4.4 Performance reporting

As defined in ANSI/ASQC Z1.15, this activity implies the performance of quality system conformity audits and reporting the results to management. Management is, of course, the normal destination of internal quality audit reports.

2.4.5 Management review

This review activity appears in several quality system standards under various titles, including Quality System Review, Review (and Evaluation) of the Quality System, Quality Control System Review, and Review and Evaluation. By their descriptions, all of these require an internal quality system audit to determine conformity and effectiveness. In most cases, the requirements imply that the audit should be carried out by management personnel. Only the Australian Standard AS 2000 actually recognizes the practical approach to meeting these requirements, i.e., that the audit be conducted by qualified, independent personnel who report their findings to management. These qualified, independent personnel would normally be quality auditors.

2.4.6 Surveillance

Normally, surveillance implies the observation of particular activities being performed as part of a production or verification process, in order to validate the action being taken, the results noted, and the decision being made.

However, the review of quality system standards undertaken by the author revealed several instances where surveillance activities have been given much broader responsibilities. In these cases the responsibility for examining and evaluating the effectiveness of the applicable quality system elements has also been included. Thus, quality audit functions have been added to the traditional surveillance activities. Where this situation arises, care must be taken to ensure that duplication of effort does not occur.

2.4.7 Product quality audit

The product quality audit is an audit or in-depth examination and evaluation of the quality system as it applies to a particular product. As such, it examines all elements of the product and their related quality system elements to evaluate the system against the referenced standards or specifications for that product.

Such an audit normally includes evaluating the ability to make valid quality decisions by those having this responsibility. This evaluation usually includes an analysis of both accept and reject decisions.

A product quality audit may include inspection or test activities as part of its data gathering methods. However, the audit should *not* be responsible for carrying out any periodic tests or inspections involved with product acceptance or rejection. At no time should the auditor be making accept/reject decisions in regard to the product itself.

2.4.8 Process quality audit

The process quality audit is an audit or in-depth examination and evaluation of the quality system as it applies to a particular process. As such, it examines all elements of the process and their related quality system elements to evaluate the system against the referenced standards or specifications for that process.

This normally includes evaluating the ability to make valid quality decisions by those having this responsibility. These decisions may be made by persons carrying out particular operations or those verifying

NEW YORK INSTITUTE
OF TECHNOLOGY LIBRARY

these operations. The audit evaluation normally includes an analysis of accept and reject decisions based on regular production inspection and test routines as well as those based on specialized periodic tests or inspections required to obtain and maintain process approval.

2.4.9 Service quality audit

The service quality audit is an audit or in-depth examination and evaluation of the quality system as it applies to a particular service. As such, it examines all elements of the service and their related quality system elements to evaluate the system against the referenced standards or specifications for that service.

Normally, this will include evaluating the ability to take actions that reflect the quality objectives of the organization and to make valid quality decisions by those having these responsibilities. In the service industry field, more of these decisions and actions will tend to be made by the people actually providing the service than by those undertaking any verification function.

2.4.10 Decision sampling

Decision sampling is a particular audit technique used to evaluate the ability of decision-making personnel to make valid quality decisions. Since both accept and reject decisions are quality-related, this technique must consider both.

2.5 Nature of the Auditor—"Who"

The nature of the auditor, i.e., the relationship between the auditor and auditee, results in the following two categories of quality audits:

1. *Internal quality audits.* These audits are conducted by auditors who are members of the auditee's organization. The position of the auditor in the organizational hiearchy does not modify this simple relationship.

2. *External quality audits.* These audits are conducted by auditors who are not members of the auditee's organization. In those cases where the auditee is also the client, the auditors are specialists from outside the organization hired to conduct an independent audit.

The implications of these two categories can be very far-reaching. Fundamentally, they hinge on the independence and objectivity of the auditor. The auditor, to be effective, must be, and must clearly be seen to be, independent of the activities being audited. Without this independence, objectivity can be jeopardized, and hence the ability of an auditor to carry out a particular type of audit can be negated.

This independence is particularly critical in conducting a suitability quality audit. It is virtually impossible for an employee of an organization to have sufficient independence from that organization's quality system to be able to assess the suitability of its quality documentation with respect to any quality system standard. Therefore, suitability quality audits must virtually all be undertaken as external quality audits.

On the other hand, conformity quality audits can be carried out by either internal or external quality auditors, as long as those individuals are not directly involved in the activity being audited.

Certain of these implications are discussed further in the chapters on planning for audits and the quality auditor.

2.6   The Method of Audit—"How"

Experience has shown there are basically two methods of conducting quality audits, namely:

1. Auditing all quality system activities at a single location during a single visit to that location

2. Auditing all activities relating to a particular element of the quality program at all locations where it applies before proceeding with activities relating to the next element or function of the program

These methods lead to the following definitions.

2.6.1   Location-oriented quality audit

The location-oriented quality audit is an audit or in-depth examination and evaluation of all the elements of a quality program having an effect at a particular location or operation within an organization. Therefore, this audit should cover the actions resulting from each applicable element of the quality program at that location and the interaction of the various elements of the program at that location.

This is a very strong technique for evaluating the actions and interactions of the various elements within the quality program. When it is used in several similar locations, differences in interpretation will quickly become evident.

However, it is difficult to apply when trying to evaluate the overall effectivity or continuity of a particular element of the program.

2.6.2   Function-oriented quality audits

A function-oriented quality audit is an audit or in-depth examination and evaluation of a particular element or function within a quality program at all the locations where it applies. By successive visits to a location, this technique can be used to evaluate a complete program.

This method is most useful when tracing the continuity of a particular element or function through all the locations where it is applicable. Therefore, its strength is the evaluation of all the actions resulting from a particular element of the program together with any interactions or reactions between locations with respect to that element.

However, it is fairly easy with this approach to miss or downplay some of the interactions between quality program elements.

2.7 Summary

This chapter has discussed the various types and variations of the quality audit. In practice, various permutations and combinations of these categories will occur. In many cases, the titles of the types will be implied rather than specifically used. In planning a quality audit, the impact of these various types must be considered regardless of the actual title used. Table 2.1 summarizes the topics discussed.

TABLE 2.1   Types of Quality Audits

Why	What	Who	How
Suitability	Quality program applicable to organization, management, product, service, process	External auditor	
Conformity	Quality system applicable to organization, management, product, service, process	External auditor Internal auditor	Location-oriented, function-oriented

Applications of
the Quality Audit

3.1 Introduction

This chapter outlines some of the applications of the quality audit. It also summarizes some of the common elements of the audit which will be developed in subsequent chapters.

The details of application will vary according to the industrial or service discipline involved. But the principles and major factors having an impact on the application of a quality audit are generic in nature and thus independent of any particular discipline being audited. These generic principles are developed in more detail later on.

First, we begin with a discussion of the three functional parties in any quality audit:

1. The party requesting the audit—the *client*

2. The party carrying out the audit—the *auditor*

3. The party being audited—the *auditee*

As indicated in Chapter 2, there are two basic types of auditors—internal, or employees of the auditee, and external, or auditors who are not members of the auditee's organization. In addition, an external auditor may be operating as an individual expert or as a member of an independent auditing organization, which then becomes one of the parties involved in the audit. In this chapter, when discussing responsibilities of the various functional parties, the term "auditor" can be taken as referring to the auditing organization as well as the individual auditor.

The relationship between the functional parties will depend on the particular application of the audit. However, in each type of audit, there are specific functions or responsibilities for each party. The parties are formally defined in ANSI/ASQC Q1-1986 as follows:

Client—is the person or organization requesting the audit.

Auditor—is the individual who carries out the audit.

Auditing Organization—is a unit or function that carries out audits through its employees.

Auditee—is the organization to be audited.

3.2 External Quality Audits

Typical external quality audits include the following:

3.2.1 Quality system certification or registration

Several national commercial and military agencies have programs that:

1. Audit an organization's quality system with respect to national or international standards
2. Then provide registration or certification of acceptable programs

In most cases, these approvals are related to particular products or product lines.

These agencies audit companies to determine if their quality systems meet the requirements of particular procurement quality standards. Thus they first determine the suitability of the quality program with respect to an applicable standard. Then, if it is acceptable, they determine the conformity of the company's operations to the quality program.

Normally, approved organizations are added to a "qualified supplier" list that usually identifies the organization's product or product line and the quality standard concerned. This approach provides better visibility for potential suppliers than the type of situation where a supplier is approved on a contract-by-contract basis.

Being included on a "quality system approval" list indicates that the organization can provide certain degrees of assurance of compliance with contractual requirements. It does not, in itself, signify approval of either the product or product line. And such a certification program does not mean a particular purchaser cannot require a higher or lower level of quality system for a particular contract. The level of system required will depend upon the application, maturity, criticality, etc., of the commodity being purchased.

In general, such certification programs recognize a company's capability of meeting requirements based on its policies and procedures and their implementation. Subsequent contractual requirements determine additions, deletions, or modifications necessary to provide the degree of quality assurance required by a particular customer on a particular order.

Some countries accept more than one standard. For example, British, Canadian, and New Zealand standards agencies have certification programs operating in the commercial field, which normally use the national quality standards as the reference standards. However, in some cases a program permits certification according to other national or international standards. Similarly, British and Canadian military agencies use NATO quality standards to register companies having acceptable quality systems for particular products or product lines.

(One important exception to the use of certification programs is comprised by U.S. defense agencies, which tend to approve organizations and facilities on a contract-by-contract basis, recognizing, of course, performance on previous contracts.)

CASE, an American commercial organization, registers supplier audits performed by its members. This listing can help customers identify suppliers that have been audited by other customers or potential customers, and may provide sufficient data to save a purchaser the cost and time involved in a further supplier audit or evaluation. The CASE listing originated with the aerospace industries as a means of reducing duplication in efforts to qualify or control suppliers in accordance with MIL-Q-9858A. It was then adopted by the nuclear industry, and is now being utilized in other commercial fields.

Programs of this nature are of great economic assistance to both large and small organizations in the areas of purchasing, marketing, and production. From a purchasing point of view, it reduces the number of supplier audits or vendor surveys that need to be carried out. From a marketing point of view, it serves to demonstrate the concern of the organization itself about quality and its ability to provide quality assurance. From a production point of view, it can reduce the interference caused by successive audits by different organizations.

This type of audit is normally initiated by the organization desiring recognition or approval of its capability to meet a particular level of a procurement quality standard, e.g., the NATO AQAP series, the British BS 5750 series, the Canadian Z299 series, etc. Therefore, the following relationships between the functional parties noted above apply:

1. *Client.*   The organization desiring certification

2. *Auditee.*   The organization desiring certification

3. *Auditing organization.* The organization granting the certification using auditors employed by itself or hired for purposes of the audit by itself

This type of audit frequently involves periodic reaudits to review any changes and confirm continued compliance with the requirements. Reaudits for change would be initiated by the audited organization seeking approval for the change. Conformance reaudits would be initiated by the certifying agency.

3.2.2 Vendor appraisal

Vendor appraisal is a technique used by many major contractors to evaluate the ability of a potential supplier to provide a particular product, service, or process. A vendor appraisal will normally cover all aspects of the potential supplier including such factors as financial stability, design capability, manufacturing capability, quality system, etc. Quality audit techniques can be useful in appraising all those areas; however, they should be mandatory for evaluating the quality system.

This type of evaluation will frequently be initiated as the result of the quality system requirements placed on the prime contractor through such standards as AQAP 1, MIL-Q-9858A, ISO 9001, ISO 9002, BS 5750, CAN-CSA-Z299, etc.

The reference standard for a vendor appraisal may be a national or international procurement quality standard or a standard developed by the contractor. The extent of the audit will depend upon the nature of the work to be undertaken, the application of the item concerned, the maturity of that item, etc.

With this type of quality audit, the following relationships between functional parties apply:

1. *Client.* The interested purchasing agent (vendee)

2. *Auditee.* The potential supplier (vendor)

3. *Auditor(s).* Member(s) of the vendee's staff or third-party auditor(s) under contract to the vendee

Once the contract has been let, periodic quality audits or appraisals may be carried out to confirm conformance to the agreed-upon quality system.

3.2.3 Vendor surveillance

Vendor surveillance, by normal definition, is the surveillance or observation of certain activities to validate the function being observed.

It is frequently used in the sense of viewing inspection or test activities at a supplier's facility rather than carrying out subsequent measurements. Vendor, or source, surveillance often provides a more effective way of ensuring that certain criteria are being met than the use of "receiving inspection" by the purchaser.

In a number of the quality standards reviewed by the author (see Appendix 1A), a requirement for determining the suitability of, and conformity to, the various controls applied to the product or process being provided is included under the term "vendor surveillance." In point of fact, however, this requirement involves an external quality audit over and above the actual surveillance or observing function.

The relationships between the three audit parties are:

1. *Client.*   The applicable purchasing agent

2. *Auditee.*   The supplier

3. *Auditor(s).*   Employee(s) of the purchasing organization or third-party auditor(s) under contract to the purchaser

3.2.4   Corporate qualified supplier lists

A particular case of vendor appraisal is the development of a corporate qualified supplier list. Those organizations approved by a corporation through vendor appraisals are listed by product line or type of business. The list then forms the basis for contacting potential suppliers with respect to future business.

Listings of this nature are also helpful in the use of corporate purchasing agreements, in which divisions or subsets of a corporation combine purchases to provide themselves with a better price break. Combining orders in this way can also enhance the assurance that the supplier maintains its goods or services at a satisfactory quality level. The larger quantities involved permit the easier introduction of statistical process control methods and other forms of statistical quality control to monitor, control, and improve quality levels.

The relationships between the three functional parties are the same as those already described for the vendor appraisal technique (Section 3.2.2). Reaudit would also be on the same basis as vendor appraisal.

3.2.5   Product liability insurance

Product liability premiums are normally based on the nature and the amount of business transacted, the risk of liability actions arising, and the probable cost resulting from any litigation that might arise.

An effective and comprehensive quality system can reduce the probability or risk of liability actions. It can also aid in any legal defense if such actions do arise. A quality audit on behalf of the insurer can determine the suitability of the quality system with respect to the product and marketplace(s) concerned and the effectiveness of that system in maintaining the quality level of the goods or services being provided.

In most cases, the reference standard would be one developed by the insurance organization. However, recognized national or international standards might also be used. The suitability and effectiveness of the quality system could have a significant impact on the size of required premium.

For this type of quality audit, the relationships between functional parties are:

1. *Client.*   The insurance company

2. *Auditee.*   The organization desiring liability insurance

3. *Auditor(s).*   Employee(s) of the insurance company or third-party auditors under contract to the insurance company

3.2.6   Regulatory controls

Regulatory agencies for the nuclear, food and drug, and insurance industries, etc., have requirements for quality systems applicable to suppliers and other organizations operating in those fields. These requirements have been introduced as a means of protecting the public. Approval based on these requirements is frequently a prerequisite to commencement of production or operation.

Agencies of this nature frequently audit organizations that fall under their jurisdiction to determine compliance with quality systems and regulations. The reference standards are the applicable agency standards and regulations.

The relationships between the concerned parties are:

1. *Client.*   The regulatory agency

2. *Auditee.*   The potential supplier or operator

3. *Auditor.*   Employee(s) of the regulatory agency or third-party auditors under contract to the agency

3.2.7   Corporate quality audits

Multidivision corporations frequently use the corporate quality audit as a means of assessing how effectively its various divisions are conforming to the quality policies of the corporation.

In most cases, these corporate audits will include:

1. A suitability quality audit to ensure that divisional quality programs conform to corporate policy

2. If the suitability audit determines conformance to corporate quality policy, a conformity quality audit to determine conformance and effectiveness of the division's operations to its own quality program

These audits are normally repeated periodically, e.g., at 1- to 3-year intervals, unless changes are introduced at either the corporate or divisional level. The periodicity depends on such factors as the marketplace, corporate quality image, criticality of the goods or services concerned, etc. If there have been no changes in either the corporate quality policy or the divisional quality program, reauditing will involve only conformity quality audits.

The relationships between the audit parties are:

1. *Client.*   The corporate head of quality

2. *Auditee.*   The division under review

3. *Auditor(s).*   Either corporate headquarters auditor(s) or independent auditor(s) under contract to the corporate headquarters

3.2.8 Product certification

Product certification is normally granted by the certifying agency after the inspection and test of initial production items. This will frequently be followed by periodic inspection and test of later samples from the production flow. The attributes or characteristics being inspected and tested, the inspection and test procedures or methods, as well as the periodicity, of the inspection and testing procedures, are normally defined in the product standard or certification agreement.

In many cases, no mention or check is made of the supplier's quality system, although this should be the means of maintaining product conformity to quality standards. However, some agencies have recognized that this approach is no longer suitable for the rapidly changing high-technology industries, if indeed it ever was for any industry. In these cases, the agency requires the manufacturer to have a quality system that provides assurance that future products will conform to the standard set by the approved units. Certification may require a quality system to meet a national or international procurement quality standard. With this approach, the reference standards become the product standard and the appropriate quality standard. This type of certification involves quality audits as parts of both the initial evaluation and subsequent reviews.

The relationships between the audit parties for this form of certification are:

1. *Client.* The certification agency, based on the application of the organization seeking certification

2. *Auditee.* The producer of the goods or services to be certified

3. *Auditor(s).* Employee(s) of the certification agency or third-party auditor(s) under contract to the agency

3.2.9 Process certification

Process certification is similar to product certification, except for the object of the review. In this case, the process may be part of a manufacturer's facility or it may be a special facility developed to provide the process as a service to a variety of potential users. Typical processes falling under this category are welding, food processing, painting, plating, heat treatment, and continuous chemical processes in the drug and biomedical field.

The reference standard will normally be a national or international standard.

The relationships between the organizations are the same as those listed for product certification in Section 3.2.8.

3.2.10 Quality system evaluation and improvement audits

These audits are usually initiated by the concerned management of an organization. The motivation may be multifaceted, but could include such factors as:

1. The desire to enter a marketplace having a requirement for a specific level of quality assurance or a specific quality assurance standard

2. An invitation to bid for a contract having specific quality system requirements

3. The desire to improve the performance or quality image of the organization

This investigative type of quality audit will normally be done by outside quality systems specialists, under contract. The specialists or consultants would be hired to evaluate the system and make any recommendations for improvement that appear necessary.

The relationships between the audit parties for this type of audit are:

1. *Client.* The management desiring to know the status of, and how to improve on, its organizational quality system

2. *Auditee.* The elements of the organization to be evaluated

3. *Auditor(s).* The staff of the third party—consultant or specialist, or independent auditor(s) under contract to the consultant or specialist

3.3 Internal Quality Audits

Internal quality audits are those conducted by auditors employed by the organization being audited. The audit may be conducted by an individual or a team selected from audit specialists, managers, executives, etc. As these individuals are involved with the quality system of the organization being audited, to varying degrees they are not normally in a sufficiently independent situation to fully assess the suitability of the quality program against some external standard. Executives, of course, would be in a position to assess such suitability relative to the organization's quality policy. Normally, internal quality audits are involved with evaluating the conformance of the various activities to the quality program and the effectiveness of the quality system.

3.3.1 Quality system audits

The internal quality system audit is primarily carried out to determine the degree of conformity of an organization's activities with its predefined quality program and the effectiveness of that program. This in turn determines whether the agreed upon policies and procedures are actually being applied effectively throughout the organization. The audit may be initiated as the result of:

1. A management desire to maintain or improve the efficiency of the organization's operations and the organization's image in the marketplace

2. A management desire to determine how disciplined and effective the organization's operations are

3. A potential contractual or certification requirement

The reference standard for this type of quality audit is the predefined, and mutually agreed upon, quality program of the organization, which defines its policies, lines of responsibility and accountability, procedures, work instructions, etc. Conformity to the documentation indicates a controlled and disciplined operation. However, the documented program may be rendered less effective through changes in requirements, techniques, technical developments, operator skills, etc., which have not been reflected in updated procedures.

Nonconformity indicates a lack of discipline and response within the system. A nonconforming observation in the course of an audit, although indicating a lack of discipline, should not automatically result in a negative observation against the operation. The situation should be investigated to determine whether the operation, procedure, or requirement should be changed. The departure may have been caused through the failure of the procedures to recognize changes or improvements in the activity or through unrealistic requirements.

For this type of audit, the audit parties are related as follows:

1. *Client.* The upper management team desiring to use the technique.

2. *Auditee(s).* The element(s) of the organization to be evaluated.

3. *Auditor(s).* Employee(s) of the organization. Their position in the organization hierarchy will depend upon the modus operandi of the organization.

3.3.2 Management review

Management reviews are key elements in managing a quality system. Such reviews make visible the interest of management in the organization's quality system, in the results of quality audits, and in the need for operational improvement where this is one of the objectives. This kind of review may involve actual audits conducted by management, either by individual managers or in teams. However, in most cases, such reviews involve analyzing audit reports, corrective action requests, and corrective actions resulting from the internal quality audits.

Management reviews are required by several procurement quality standards. However, most of those standards ignore (even if they do not negate) the fact that the actual auditing activities used to provide the data for such reviews are usually delegated by managers to others.

As with all forms of internal quality audits, the reference standards are the policies, lines of responsibility and accountability, procedures, work instructions, etc., forming the documentation of the organization's quality program.

The audit parties are related as follows for this type of quality audit:

1. *Client.* The executive of the organization

2. *Auditee(s).* The applicable operating element(s) of the organization

3. *Auditor(s).* The manager(s) assigned to the audit or their delegates

3.3.3 Performance reviews

"Performance review" is a term used in some standards. It is applied to management review activities and also to the analysis of the findings and subsequent actions resulting from internal quality audits. Such reviews compare the findings of the audits with the objectives of the organization, e.g., with the organization's policies, procedures, etc.

At least one procurement quality standard specifically requires a performance review.

The reference standards and relationships between functional parties are the same as those for management reviews (Section 3.3.2).

3.3.4 Product quality audit

A product quality audit is basically a quality system audit applied to the system elements related to a particular product. This may involve special periodic tests or evaluation techniques carried out on the product to ensure that there has been no degradation over a period of time. It should be noted that these tests are performed to provide data on the effectiveness of the control system and do not form part of the product accept/reject decision cycle. They should not be confused with periodic tests or inspections that do form part of the accept/reject decision sequence for the product or batches of product.

The reference standards for a product quality audit are the product quality program and the product performance specification.

The relationships between the audit parties are the same as that shown for quality system audits, Section 3.3.1.

3.3.5 Process quality audit

Process quality audits are similar to product quality audits except for the object of the audit. Similarly, the reference standards and audit party relationships are the same as in those described for product quality audits (Section 3.3.4).

3.3.6 Service quality audits

Service quality audits are similar in all respects to product quality audits (Section 3.3.4) except for the object of the audit. In conducting this type of audit, it will frequently be necessary for the auditor to sample the service being provided. Decision sampling (Section 3.3.7) forms a significant tool in the evaluation of that service.

3.3.7 Decision sampling

Decision sampling is an audit technique used to validate the ability of individuals to make valid quality decisions. Basically, inspection, test, and other verification personnel are paid to make valid quality decisions based on the results of their inspection, test, or verification activities. The actual measurement activity is a means of providing data on which to base a decision and not an end in itself. This technique is discussed in more detail in Section 10.2, Decision Sampling.

The reference standards will depend on the operation being evaluated. The relationships between the audit parties are:

1. *Client.*   The management initiating the review

2. *Auditee(s).*   The decision maker(s)

3. *Auditor(s).*   The employee(s) using the technique

3.3.8   Data processing quality audit

Data processing quality audits fall into two basic categories:

1. A specific form of process control quality audit used to determine if the programmers are following the approved software quality program, i.e., its procedures, operating instructions, etc.

2. A specific form of decision sampling used to determine the ability of the data entry personnel to follow the software quality program and to enter and verify the accuracy of the data entered. This becomes a particular application of decision sampling.

The reference standards are the documentation defining the objectives, techniques, and operations of individuals concerned with software and data processing.

The relationships between audit parties are similar to those described for process quality audits (Section 3.3.5).

3.3.9   Customer service quality audits

Customer service quality audits are a subset of service quality audits (Section 3.3.6). Normally they cover two aspects of customer relations:

1. An evaluation of whether or not the customer is truly being provided with the intended service, as documented in the agreed-upon quality program. This will frequently require the auditor to make use of the service being provided.

2. An evaluation of how customer problems are answered, handled, analyzed, and reported in accordance with the agreed-upon quality program.

The reference standards used are the applicable elements of the quality program. The relationships between audit parties are the same as those described for process quality audits (Section 3.3.5).

3.4 Conclusion

The description of typical applications of the quality audit show massive similarities between the various types. These similarities permit a generic approach to be taken in the preparation of quality audit standards, guidelines, etc., as well as in our discussion of quality audits throughout much of the remainder of this text. In turn this generic approach makes it possible to divide the study of quality audits into the following major topics:

1. Initiating the quality audit (Chapter 5)

2. Planning the quality audit (Chapter 6)

3. Implementing the quality audit (Chapter 11)

4. Measuring the effectiveness of the quality audit (Chapter 15)

In the chapters that follow, each of these major headings will be further subdivided into more detailed activities based on the different types of quality audits already noted and their complexities.

4

Managing the Quality Audit

4.1 Introduction

There are three different viewpoints involved in managing a quality audit, resulting from the three parties concerned: the client, the auditee, and the auditing organization [auditor(s)]. Regardless of the relationships between them, each of these parties have responsibilities that must be addressed.

The management activities themselves subdivide into two segments:

1. Administrative
2. Technological

4.2 Administrative Management

Each party has the normal administrative functions that are involved with management responsibility. These general skills are covered by many texts and are largely outside the scope of this work. Suffice it to say at this point that the style of general management can have a significant impact on the effectiveness of a quality audit. Maximum value from the audit and ease of obtaining data will occur in those organizations that use a cooperative form of management. A more adversarial form will make it extremely difficult, if not impossible, to obtain all the necessary information during the audit. With the interactions between auditors, lead auditors, audit supervisors, etc., it is strongly recommended that a cooperative style of management also be used within the auditing organization.

4.3 Technological Management

The managerial activities directly associated with the quality audit could be referred to as the "technological mode" of management.

These activities involve each of the audit parties, which have specific responsibilities regardless of the interrelationships between them.

The purpose or objective of a quality audit can have a direct impact on the environment surrounding the audit. While every endeavor must be made to maintain a cooperative climate, occasions can arise where the atmosphere becomes adversarial. An example of this latter condition would be an audit being carried out to obtain evidence on behalf of the plaintiff in a liability investigation or litigation. Even in these situations, the auditors should try to maintain a constructive approach. A quality audit should be a constructive experience for each party involved in it. Every effort should be made to maintain a positive working environment.

The technological aspects of managing a quality audit follow the traditional management cycle, with the resulting inward spiral due to continuous improvement of the process:

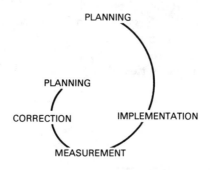

PLANNING

PLANNING

CORRECTION   IMPLEMENTATION

MEASUREMENT

TYPICAL MANAGEMENT SPIRAL

This cycle applies whether one is dealing with an external or internal quality audit. The participants may have different titles and relationships to each other depending on the type of audit; however, their responsibilities, functions, etc., remain fundamentally the same.

4.4 Function Identification

In order to efficiently manage the functions of the quality audit, each activity to be performed must be known, i.e., each major function and its subsets must be identified. The quality audit is a self-contained function for most of its activities. However, as with all operating functions, some aspects of the work are dependent on other bodies performing their tasks correctly and at the appointed times. As a self-contained function, the quality audit can be examined using a "function tree." This technique has been shown to be one of the strongest function analysis tools available.

4.5 Basic Function Tree

The function tree is an analytical tool for examining major functions within an organization. It takes a major function (function A in Figure 4.1, for example), then subdivides that into the second-level functions contributing to the major activity (functions AA, AB, and AC in Figure 4.1). This sequence is then repeated until all contributing activities have been identified. These are represented by the third, fourth, and fifth levels or columns in Figure 4.1. The number of levels will depend on the complexity of the function being analyzed. Once the tree has been developed, detailed analysis starts with the functions in the right-hand column and progresses toward the left. This technique also serves to make common activities clearly visible.

As well as an aid in identifying the various activities that comprise a quality audit, the function tree is a powerful tool for planning a quality audit. Therefore, to demonstrate this, in the following paragraphs and chapters I will use the function tree to identify the various activities or functions leading to a successful quality audit. The number of activities and levels concerned necessitate subdividing the function tree into several illustrations instead of showing it as a single large diagram. But the relationship between the various levels will be clear through the use of a numbering system similar to that used in Figure 4.1.

4.6 Function Tree—Quality Audit

The prime function—managing the quality audit—subdivides into the following second-level functions, combining the prime audit functions and their evaluation:

1. Initiation of the quality audit (discussed in Chapter 5)
2. Planning the quality audit (discussed in Chapter 6)
3. Implementation of the quality audit (discussed in Chapter 11)
4. Interpretation or analysis of the quality audit (discussed in Chapter 12)
5. Reporting the quality audit results (discussed in Chapter 14)
6. Corrective action requests and followup with the auditee (discussed in Chapter 13)
7. Measuring the effectiveness of the quality audit (discussed in Chapter 15)

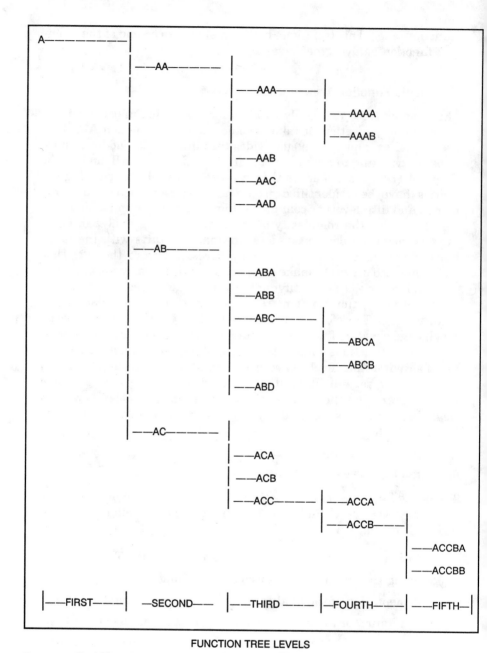

FUNCTION TREE LEVELS

Figure 4.1 Basic function tree.

8. Corrective action and followup with the auditors (discussed in Chapter 16)

These first and second levels of the quality audit functions are shown diagrammatically in Figure 4.2 as the first two levels of the quality audit function tree. Each of the second-level functions can be subdivided in turn into their respective subsidiary functions, as illustrated in the figures in later chapters.

4.7 Audit Management Documentation

4.7.1 General documentation

There are several aspects to the responsibilities of each party in the quality audit that should be documented to ensure adequate control of the activities. Since in different situations an organization is liable to have to fulfill the obligations of any of the three functional parties, the wise organization will have a set of procedures covering each category—client, auditee, and auditing organization. This documentation may take the form of written procedures, operating or work instructions, etc. But whatever the form, it should form an integral part of the quality system.

4.7.2 Client documentation

On the surface, the responsibilities of the client in initiating the quality audit are relatively simple:

1. Stating the purpose of the audit
2. Identifying the reference standard to be used by the auditor

However, there are pitfalls to be avoided in both areas. As indicated in Chapter 2, there are many different purposes or objectives for a quality audit. A number of these can apply within a given organization in different situations. Therefore, the procedures should show how to select the right type of quality audit for the particular situation. They should be written in a manner that ensures continuity of action from audit to audit, within a given type. However, they should not inhibit the creative spirit of the personnel concerned, but instead leave them feeling free to develop new and improved audit techniques and applications.

Similarly, in the field of reference standards, the procedures should maintain continuity without inhibiting the search for new and more appropriate standards. The documentation should guide in the selection of industrial, national, and international standards covering

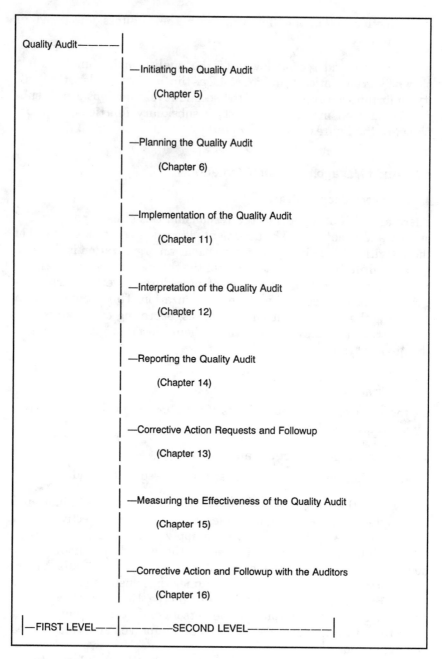

Figure 4.2   Function tree for managing the quality audit.

quality systems, quality audits, quality and process statistics, regulations, products, services, or processes, etc. As companies develop international markets, the number of standards, regulations, etc., they must be concerned with increases rapidly. Without proper documentation, the wrong reference standard can very easily be selected.

4.7.3 Auditee documentation

The documentation for the auditee describes that party's activities during three phases of an audit:

1. What to do when notified that a quality audit is to be conducted
2. What to do during the quality audit itself
3. What to do with the results of the quality audit

Each of these areas must be addressed to ensure that all members of the auditee's organization are aware of their responsibilities during each phase of the audit, as well as which actions are permitted and which are not. Questions will also arise where special guidance will be required. These include:

1. What to do when notified that a quality audit is to be conducted:
   a. Has the quality system been designed to meet the requirements of the proposed audit?
   b. When will key members of the organization be available during the audit's time frame?
   c. Is the quality system fully documented?
   d. Does the available objective evidence demonstrate compliance with the quality program?
   e. Are there any restrictions on access to facilities or records for national or industrial security reasons, safety reasons, etc.?
   f. What office space is available for use by the auditor(s)?
2. What to do during the quality audit:
   a. Who will attend the meetings involved with the audit?
   b. Who will escort the external auditors while they are in the plant?
   c. Should internal auditors be escorted at any time?
   d. What access to other personnel will the auditors have?
   e. What freedom will the auditors have in duplicating any records?
   f. What reports are to be prepared during or on completion of the audit?
3. What to do with the results of the audit:

    a. Who are responsible for taking action on verbal or written observations or comments made during or after the audit?

    b. What actions, if any, are to be taken as the result of verbal comments made during the audit?

The answers to most of the above queries will depend on the modus operandi of the auditee. There is no single panacea for all these situations. Suffice it to say that each of them must be addressed and answered in a way that will be understood by all concerned. Where the answers will inhibit or restrict the auditors, it is essential that the auditors be fully apprised of the problems as soon as possible, preferably at the time of audit initiation. In some cases, the actions required to remove an unacceptable inhibition may have a serious impact on an audit's scheduling, e.g., obtaining necessary government security clearances.

4.7.4 Auditing organization documentation

The documentation required by an auditing organization should cover at least the following topics:

1. The requirements for following the ethics of a quality audit (Section 6.2.5), in particular the clauses affecting auditor independence, confidentiality, and interactions with staff members of the auditee.

2. The preparation of various types of working papers needed: instructions, check lists, memory prompters, etc.

3. Guidance on the interpretation of the various quality system and quality audit standards to ensure a common viewpoint and thus continuity from auditor to auditor. Standard sheets should be developed for each system standard as it is identified. Each new work sheet, developed for determining the suitability of a particular quality program, should be added to the existing library.

4. Guidance on the best approach to questioning personnel at the organization being audited, as well as on other points.

5. The methods to be used in recording observations, comments, etc., on the quality system being audited.

6. Information on the various sampling and interpretive statistical techniques to be used.

7. Guidelines on how to prepare the quality audit reports, and the forms necessary for doing so (this report is the end product of the audit).

8. Guidelines on how to determine which activities require corrective action, how to request it, and how to follow it up.

The auditing organization documentation should be in a form that can be provided to each auditor as a reference instruction. Although these are a type of procedures, I favor the term "Quality Audit Instructions." Under this heading they can readily be used to clarify sections of general procedures for use by the auditors.

4.7.5 Conclusions

The actions and reactions resulting from interpersonal relationships among those participating in a quality audit have a significant impact on the effectiveness of the audit. In fact, the success of the audit depends on the cooperation of all three functional parties. For this reason, I believe it is important to document the actions of each party, so that each individual involved is absolutely clear about his or her responsibilities, accountabilities, actions, and reactions. Although this knowledge is of paramount importance, it must be recognized that every situation cannot be covered in the documentation. Thus the instructions should be phrased in such a way as to provide general principles and guidance.

5

Initiating
the Quality Audit

5.1 Introduction or Initiation of the Quality Audit

Once the client has expressed a desire to use a quality audit, the three parties concerned, i.e., the client, the auditee, and the auditing organization, should have a meeting of minds on initiating the audit. Any detailed planning, hiring or training of auditors, obtaining resources, etc., should be delayed until afterward. At that meeting, a consensus should be reached on the purpose or objective of the proposed audit, the benchmark or reference standard against which documentation and activities will be compared, the auditors to be used, and the sequence and schedule for the major events associated with the audit. The function tree for the activities related to initiating an audit is illustrated in Figure 5.1.

5.2 Initiation Participants—External Quality Audits

The client, as the initiator of the quality audit, should be represented by the senior member of the function requesting the audit. If this representative is not the function manager, the representative must be able to speak for, and take decisions on behalf of, the manager. Potential sources for this task are indicated under the various types of external quality audits described in Chapter 2.

Similarly, the representative for the auditee should be that organization's responsible manager, accompanied by a senior member of the organization's quality function.

Likewise, the auditing organization should be represented by its

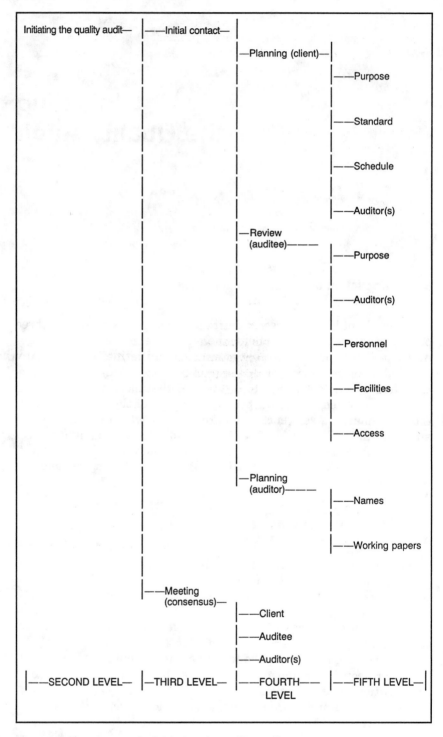

Figure 5.1 Function tree for initiating the quality audit.

head of the auditing activity. This individual may be a member of the client's organization or an independent third-party auditor, but *not* a member of the auditee's organization.

5.3   Initiation Participants—Internal Quality Audits

The same general ground rules and responsibilities apply for the initiation of an internal quality audit as for an external one. The major difference is in the relationships between the attendees, as shown below:

1. *Client.*   The Chief Executive Officer (CEO) or the head of the element desiring the quality audit, e.g., Subsidiary President, Group Manager, Division Manager, etc.

2. *Auditee.*   The manager(s) of the functions to be audited.

3. *Auditing organization.*   The manager given the responsibility for the quality audit. The manager of the audit activity should be independent of the functions being audited.

The manager in charge of the organization's quality function may, or may not, be assigned the responsibility for managing the audit. If the line functions are responsible for demonstrating that they have carried out their assigned tasks correctly, i.e., if they are responsible for an inspection and test function on which quality and acceptability decisions are based, the manager of the quality function could manage the audit. But if test and inspection is carried out by the quality function, then the audit function should be under an independent manager unless the quality function has an audit supervisor on at least the same management level as the supervisors of the inspection and test functions. The author introduced an internal quality audit in this latter environment.

5.4   Planning the Initiation

5.4.1   Client activities

5.4.1.1   Client responsibilities   The client's approach to the quality audit basically determines the climate or working environment for the quality audit. This impact commences when the first contact is made with the auditee.

The client has the following fundamental responsibilities:

1. Initiating the audit

2. Defining the reference standard(s)

3. Receiving the quality audit report

4. Determining any followup action required

5.4.1.2 Client initiation of the audit   The client initiates the audit by bringing the three functional parties—client, auditing organization, and auditee—into contact to develop a *mutual understanding and agreement* on:

1. The purpose of the quality audit

2. The reference standards that will serve as benchmarks

3. The scheduling or timing of the audit

4. Guidelines on how each party can abstain from any undue interference with the auditing

This understanding and agreement can best be reached at a meeting between the three parties. The correspondence setting up the meeting should briefly state the purpose of the audit and the reference documents. It should also indicate a date by which the audit activities are to be completed. This will enable both the auditee and the auditing organization to prepare for the meeting.

The first two of the four points listed above are the prerogatives of the client to determine. However, both the purpose of the audit and the reason for the selection of the particular reference standards must be clear to all the parties. The purpose or objective of the audit should be defined so that the auditee will understand it to be a constructive program and not just a "fishing" operation that could lead to punitive action. The benefits that can be expected from the audit by both the client and the auditee should be clearly outlined at the meeting.

As discussed in Chapter 3, the reference standards can take many forms, e.g., national, international, or corporate quality system standards; product standards; government regulations; auditee or corporate quality policies, etc.

The scheduling or timing of a quality audit can easily get the project off to a bad start. The discussion on scheduling should be based on a broad time frame stated in the letter setting up the initial meeting; this will help the three functional parties to develop a schedule that is *mutually* acceptable. The need for this mutual development of the schedule is frequently overlooked. I have seen too many incidents where an organization is notified that an audit team will arrive on a certain date, requiring certain facilities and personnel to be provided by the auditee. The auditee resents this autocratic approach, which can then lead to an adversarial atmosphere during the actual audit.

The need for the client and auditee to abstain from any interference with the auditing activity must be understood and agreed to by all parties. This abstinence is required to enable the auditor to be responsible for observing and interpreting all of the findings during the audit. Interference could lead to potentially pertinent observations being overlooked or misinterpreted.

At the initial meeting, the client can also state its desires on distribution of the audit report and what actions it expects as the result of the report. This part of the discussion should clearly indicate the amount of information to be provided verbally by the auditors during the audit.

5.4.2 Auditee activities

5.4.2.1 Auditee responsibilities   The auditee has the following responsibilities with respect to the preparations for, and implementation of, the quality audit:

1. Agreeing on the purpose or objective of the audit and accepting this and the defined reference documents
2. Agreeing on the acceptability of the proposed auditor(s)
3. Appointing responsible individual(s) to work with the auditor(s)
4. Providing a working area and facilities for the auditor(s)
5. Ensuring auditor access to the necessary facilities, objective evidence, etc.
6. Attending specific meetings with the auditor(s)
7. Reviewing the audit findings to ensure agreement with the facts

5.4.2.2 Auditee initiation activities   The auditee's representative should be in a position to discuss the acceptability of, the purpose of, and the reference documentation for the proposed quality audit.

Accessibility to facilities, objective evidence, special processes, etc., may be restricted owing to numerous factors, including:

1. Restrictions on particular facts due to government or industrial security needs, proprietary information relating to other customers, etc.
2. An apparent lack of objectivity by the proposed auditor(s) based on some stated reason, e.g., previous employment, financial holdings, involvement in development activities, etc.
3. Safety regulations
4. Conflicting schedules between the auditee's operations and the audit

These problems must be resolved to the mutual satisfaction of all three parties. During the initial meeting, the schedule for the audit should be defined specifically, so that all participants are working toward the same start date.

5.4.3 Auditing organization

The representative of the auditing organization at the initial meeting should have available for discussion:

1. The names of the auditor(s) to be involved. If more than one is required, the lead auditor should be identified.

2. The time required to prepare any special working papers required for the audit. *Note:* "Working papers" are defined in American and Canadian quality audit standards as instructions, check lists, reporting forms, etc.

3. Recommendations on where a suitability audit should be carried out, if required, e.g., at the auditor's base, in the auditee's facility, etc.

4. Details of the facilities and support that can be expected from the auditee.

Once the general logistics for the up-coming quality audit have been finalized at the preliminary meeting, the lead auditor, assisted by the other members of the audit team, will carry out the detailed preparations we will discuss in later chapters of this book.

The same general pattern should be followed for either an external or an internal quality audit. The major difference between the two to be noted at this point is that an internal audit will seldom be carried out by a team. Undoubtedly, it will involve more than one auditor. However, each auditor will be responsible for particular elements of the system being monitored. To maintain versatility, auditors should be rotated through different elements wherever possible.

5.5 Conclusion

As stated in the introduction, a quality audit is initiated through a preliminary "meeting of the minds" on the part of those to be involved in it. This can be achieved through a physical meeting, teleconferencing, or one of the other techniques used to bring individuals together.

As the result of the initiation meeting, consensus should be reached on *what* activities are to be undertaken, *where* these activities will be carried out, *when* they will happen, and *who* will be involved. This *mutual understanding and agreement* will enable the auditee and the auditing organization to carry out their activities in the most effective manner.

More detailed aspects of planning for a quality audit, implementing it, measuring the findings, and following up on corrective actions will be discussed in later chapters.

6

Planning the Quality Audit

6.1 General Planning

The planning activities for a quality audit by the auditing organization cover a wide gamut of functions involving actions and reactions from the three parties that will be involved in the audit—the client, the auditee, and the auditing organization itself. These activities include the following (they and their subsets are shown in Figures 6.1*a* and 6.1*b*):

1. Determining implications of the quality audit (discussed in Section 6.2)

2. Understanding the resources needed for the audit (discussed in Section 6.3)

3. Scheduling the audit (discussed in Section 6.4)

4. Sequencing the quality audit functions (discussed in Section 6.5)

5. Preparing or gathering working papers for the audit (discussed in Section 6.6)

Note: "Working papers" is used in the sense of paragraph 5.2.4 from ANSI/ASQC Q1-1980: "These are all documents required for an effective and orderly execution of the plan."

6. Determining the sampling procedures to be used in the audit (discussed in Section 6.7)

7. Interpreting the quality audit observations (discussed in Chapter 12)

8. Reporting the results of the audit (discussed in Chapter 14)

9. Requesting and following up on corrective action (discussed in Chapter 13)

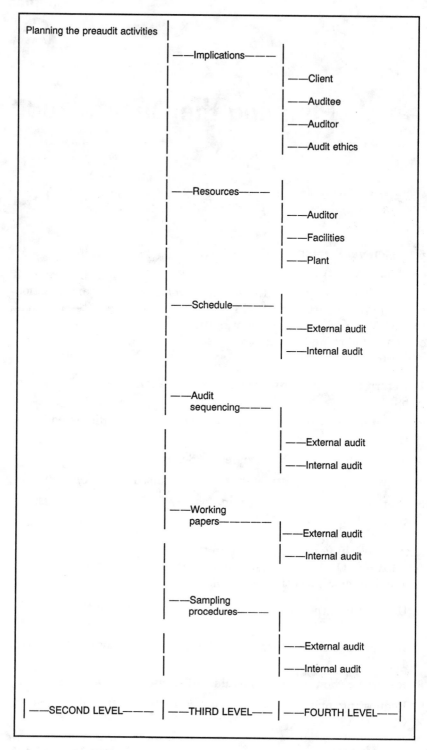

Figure 6.1a Function tree for quality audit planning—preaudit activities.

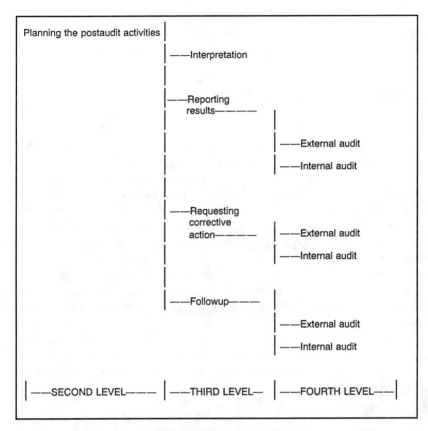

Figure 6.1*b* Function tree for quality audit planning—postaudit activities.

As the principles applying to each of the above activities can be applied to virtually any type of audit to be undertaken in any kind of industrial or service environment, the following discussions and examples will be couched in generic terms. If a particular point has connotations for a more specific situation, it will be identified as such, along with the reasons for the specificity.

The planning activities may be undertaken by the head of the auditing organization or by senior members of that organization's staff. The overall planning should undoubtedly be the responsibility of the head of the organization. The more detailed planning with respect to particular standards, auditing procedures, etc., should be prepared by the individual responsible for the particular audit being planned, or by the lead auditor if a team is involved. The degree of delegation will depend on the size of the organization, the abilities of its staff to handle the delegated activities, the criticality of the audit, the modus operandi of the auditing organization, etc. In general, however, the delegation of certain segments of the planning activities to lower-level auditors will serve as excellent training in their career development.

6.2   Implications of the Quality Audit

6.2.1   General implications

Regardless of the type or application of a quality audit, it is carried out to examine and evaluate the working quality system of the auditee. This implies that there *is* a system and that it is functioning and results can be measured. If the system is to be evaluated, it is essential that there be some form of documentation outlining the objectives for each of the various operating elements. This implies some form of a documented quality program to serve as the object of a suitability audit and the benchmark for the conformity audit. The documents should be in the form that most effectively communicates the desired message to the intended readers. Certain procurement quality standards are very specific on the form these documents should take; others are less explicit.

Regardless of the requirements of any standard, it is essential that formal documentation of the quality system should exist in a form which ensures that all personnel can clearly understand their responsibilities and the lines of accountability.

6.2.2   Implications for the client

The major implication of a quality audit as far as the client is concerned is the need to know exactly what level of quality system complexity will be required to satisfy its quality needs in regard to product or service. The need arises from the type of quality audit desired, as outlined in Chapter 2. The selection of complexity level is extremely critical when considering the value to be obtained for the cost of the program. If too complex a quality system is specified, the expenditures, by the auditee and the auditing organization alike, in achieving and verifying it may greatly exceed the value to either the client or the auditee. On the other hand, too simple a system may not provide the necessary assurance that the product or service will meet the needs of the customer.

On receipt of the audit report, the final decisions on how the audit is to be used, how the report is to be distributed, if any further action is required, etc., rest with the client. There are basically three decisions in this are available to the client:

1. The quality system meets the requirements of the reference standard.

2. The quality system does *not* meet the requirements of the reference standard. However, with certain specific corrective actions it can be made acceptable.

3. The quality system does not meet the requirements and does not justify requesting detailed corrective action.

The choice of which decision is applicable will depend on the needs of the client and on the nature of the client/auditee relationship. A decision that a quality system is inadequate is particularly sensitive to these factors.

As the customer of the quality audit, the client owns the data derived from the audit, the analyses carried out by the auditor, and the final audit report, and as proprietary owner may use them in any manner it desires. For this reason it is extremely important that the policies and ground rules relating to how observations, analyses, and reports are to be made are clearly defined to both the auditor and the auditee prior to the audit. This responsibility lies with the client, and is particularly important in cases where potential suppliers are being assessed or where approval of products, services, or processes is concerned.

6.2.3 Implications for the auditee

The major implications relating to a quality audit are threefold for the auditee:

1. Having a working quality system compatible with the reference standard

2. Having the quality system described in documentation available to, and understood by, all applicable personnel

3. Having objective evidence demonstrating compliance of the system with the documentation

These factors would appear to be normal requirements for a well-managed organization. However, many organizations, large and small, will not document their operations as they feel doing so would inhibit the personnel concerned. They fail to see that the documents do not restrict creative thinking about process improvement, but rather provide a means of measuring improvement with respect to a known base. They also provide continuity from person to person on the same function, as well as offering new members of the team a reference to use while learning their jobs.

6.2.4   Implications for the auditor

Each new audit increases the challenge of maintaining confidentiality on improvements that can be made. As auditors gain auditing experience, they also gain experience and knowledge about different approaches to the development of systems and processes. This increased knowledge and experience should improve their ability to assess different approaches to analyzing basic operations of organizations to be audited or to frame effective corrective action proposals, *not* their ability to provide improved problem-solving methods for the auditee.

6.2.5   Quality audit ethics

Ethics must be a significant factor in planning and conducting a quality audit. A code of ethics can affect the eligibility of the auditing organization and the individual auditors just as much as the type of audit to be undertaken.

A code of ethics for quality auditors has been under review by the Quality Audit Technical Committee of the American Society for Quality Control for some years. The code suggested by this author is given in Appendix 6A.

A code is particularly important in helping auditors attain an independent objectivity in their observations, their evaluations, and their subsequent analyses. This objectivity must not only be there in practice, it must also be clearly evident to all the parties associated with a particular audit, as well as to others viewing the activities associated with it either directly or indirectly. The auditor is in a position analogous to Caesar's wife, who had not only to be pure but to be visibly pure to all others.

A number of different associations between the auditor and the auditee can contribute to the existence, or at least the appearance, of a lack of objectivity or independence on the part of an external auditor:

1. Previous employment of the auditor by the auditee, regardless of the reason for separation.

2. Previous employment of the auditor by a major competitor of the auditee, regardless of the reason for separation.

3. Holding of significant amounts of stocks or bonds by the auditor in the auditee's business or that of a major competitor of the auditee.

4. Previous association of the auditor with the development of the

quality system under evaluation to meet a particular quality standard. (This would preclude an auditor from evaluating the suitability of the system against that reference standard. However, such a relationship would not preclude the auditor from evaluating the compliance of the system with the documented quality program.)

Needless to add, auditors should always avoid even the hint of such improprieties.

6.3   Resources Needed for a Quality Audit

6.3.1   General

In examining the various resources required for a quality audit, the auditor as a resource justifies a complete chapter in itself, whereas the other resources needed can readily be covered as sections in this chapter. The details concerning resources will depend largely on the type of audit and the relationship between the auditee and the auditor, i.e., whether the audit is to be internal or external in nature.

6.3.2   Personnel

The quality auditor requires a blend of skills, dichotomies, and attitudes developed through the quality and the technological or service disciplines encountered during his or her career. These are discussed in detail in Chapter 7, "The Quality Auditor."

Staff of the auditee will be involved in both internal and external quality audits, to differing degrees.

6.3.2.1 Participation by auditee personnel—external quality audits During an external quality audit, members of the auditee's staff will be required for a number of activities, including:

1. Accompany the auditors throughout their audit activities. This is to provide guidance and assistance
2. In each of the work areas visited by the auditors, answer questions and demonstrate both the work activities as they relate to the organization's quality system and the objective evidence that the quality standards are being adhered to by the system element being reviewed
3. Attend the preaudit and postaudit meetings
4. Perhaps operate the product or service being audited
5. Perhaps operate specialized verification facilities such as high-technology metrology devices, environmental facilities, measure-

ment equipment, etc., that may lie outside the skill of the auditors to operate

6.3.2.2 Participation by auditee personnel—internal quality audits   The support required from auditee staff for an internal quality audit is virtually the same as that needed for an external one, except that it should not be necessary for someone to accompany the auditors throughout their audit activities.

6.3.3 Office facilities

Office space is required for both external and internal audits; in the first case temporarily and in the latter on a permanent basis.

6.3.3.1 Office services—external quality audits   The primary need external auditors have for office space is for a place where they can have privacy to review their observations and plan any variations in the sequence or other aspects of the audit. External auditors also require access to a telephone, although it need not be located in their office area.

In the interests of confidentiality, external auditors normally prepare reports and corrective action requests at their own home office. Thus they should not require the auditee to provide typing or word processing facilities. However, the same need for confidentiality of observations, comments, etc., means the auditee *should* provide some facility for the safe storage of records.

6.3.3.2 Office services—internal quality audits   Internal auditors have the same needs for privacy and the ability to store confidential information safely as do external auditors. Therefore, the office space allocated must recognize this. Even in an open office site, judicious allocation of space can provide the necessary environment.

Moreover, internal auditors will require technical support facilities beyond telephones—computer terminals or personal computers for secure analysis, for example. Since reports are the end product of the audit itself and all future actions or reactions will be based on them, auditors should be encouraged to use dictating equipment, which, properly used, can speed the preparation and transcription of reports. A word processing system should also be available in the auditors' work area to assist them in the rapid preparation of special action reports such as corrective action requests. Documents printed out from word processors are more readable than those that are handwritten. Ready access to copiers is another key factor in timely reports.

As noted above, secure storage space is required to keep the reports, observations, etc., resulting from the audits confidential. If the actual paper documentation is retained, the space requirements will be substantial. But if a secure word processor or computer is available, then paper copies of only active documents need be retained. All other documentation can be stored in the word processor or the computer, and the paper documents destroyed once the specific audit activities with which they are concerned are completed.

In many audits there will also be performance charts showing the various trends. These should be clearly visible for everyone to see. In some cases, there may duplicate charts in the work area being examined and in the audit area. But every chart should be situated so that all the interested parties have ready access to it.

6.3.4 Plant facilities

Most quality system standards require the contractor (the auditee) to make available to the customer's representatives certain verification facilities for use in verifying the contractor's data. Under this heading would be included such items as special metrology devices, standards for calibration of metrology items, specialized facilities for verifying vibration levels, environmental conditions, tensile strengths of products, etc. Access to facilities of this nature would be required by both external and internal auditors. Decision sampling techniques could reduce the need for access to some of them.

6.4 Scheduling

6.4.1 Scheduling—external quality audits

The sensitivity in scheduling external quality audits was already discussed in Chapter 5, "Initiating the Quality Audit." The need for adhering to the "mutual agreement" approach in setting the audit dates cannot be stressed too strongly. An autocratic approach of unilaterally setting the dates for an audit has resulted in more resentment over audits than any other single feature.

The length of time needed for an audit will, of course, depend on the complexity of the quality system being audited, the criticality of the items under review, etc. Care should be taken to allow sufficient time to cover any delays that might occur. It is probably best for the auditors to quote a duration period in terms of the most likely length of time the audit will take, along with suggested minimum and maximum duration parameters. Finishing early will seldom cause a problem. However, extending the audit period can definitely lead to problems.

The success of an audit depends on cooperation between the auditor and the auditee. This can only be achieved and maintained through mutual respect for the needs of each other. As the audit progresses, it may be necessary to adjust already scheduled activities to accommodate unforeseen requirements on the part of individuals or facilities involved in the audit. This type of flexibility within the overall time frame can greatly help in maintaining an optimal atmosphere of co-operation. It should always be recognized that the mere presence of an outside body in a working area can cause some disruption to the workers there, with a possible loss of worker care and productivity.

6.4.2 Scheduling—internal quality audit

Internal quality audits are, in general, recurring activities with some predetermined periodicity. Thus the time frame approach of the external quality audit does not apply.

The scheduling of internal audits is not as sensitive as that of external audits, since both auditee and auditor are working for the same organization and hopefully toward common objectives. However, it is important that all parties are made aware of the repetitive nature of an internal audit. Periodicity is normally given in terms of so many complete audits per unit of time, i.e., per day, week, month, or year, which may vary from work area to work area or work function to work function. Initially it depends on the criticality of the elements concerned and the modus operandi of the organization. Later on it will depend upon the performance of the individual work areas. Some quality system standards place a maximum period over which a complete system audit can be spread.

Although the periodicity of internal audits is usually known, the timing of actual audit visits is seldom published. Every endeavor should be made to introduce a degree of randomness, so as to prevent a pattern developing which may bias the audit results. Industrial engineering random-time tables are a useful aid toward this end.

Periodicity should be published, but not cast in concrete. As data is derived, there should be allowances for increasing or decreasing the length of the period between audits. The rules governing such changes should be known to both the auditor and the auditee.

6.5 Sequencing the Quality Audit

Sequencing the audit activities requires a thorough analysis of the audit's requirements and the operating system being audited. Suitability audits will normally work through the reference standard on a paragraph-by-paragraph basis. The conformity quality audit requires a detailed evaluation of the operating system involved using one of the planning tools outlined in Chapter 8. Once the system has been analyzed, the applicable documentation should be noted for each function. This should reveal any gaps in that documentation or the presence of other applicable documentation not in use. Once this has been done, a decision can be made on whether to undertake the audit in a location- or function-oriented mode.

6.6   Working Papers

Working papers are covered in Chapter 9. They are defined in audit standards as all the documentation required for the audit activity, i.e., procedures, working instructions, check sheets, memory prompters, records, etc. Suffice it at this point to highlight certain key factors:

1. Working papers include memory prompters for the auditor to ensure no elements are missed by the review and evaluation.
2. Working papers should not be mere checkoff sheets on which to note the presence or absence of predetermined elements of a quality system. They should allow, at least, for the grading of the effectiveness of those elements.
3. Working papers include detailed records of what was audited and where the activities of the audit were carried out, plus the findings for each audit activity.

The general principles for the various working papers are the same regardless of the type of quality audit or who the auditors are. But each type of audit will have its own distinctive papers. External auditors will gradually develop a library of documents as they are exposed to different system and product standards. Suitability audits carried out in light of given standards should use the same papers regardless of the auditee.

6.7   Statistical Techniques

6.7.1   Sampling procedures

All quality audits require sampling techniques, in some form or another. The following fundamental principle must apply at all times:

> All sampling procedures used in a quality audit shall have a sound statistical foundation.

Without this principle, the soundness of any audit can be questioned. Sampling procedures are discussed in detail in Chapter 10.

6.7.2   Evaluation statistics

Wherever possible, the results of a quality audit should be quantified. Without quantification, it is extremely difficult to measure trends between audits or to accurately compare performances. Various statistical techniques will be used in the evaluation of the audit observations. These are discussed in Chapter 11, "Analysis of the Quality Audit."

Appendix 6A
Code of Ethics for Quality Auditors

6A.1 Fundamental Principles

In order to uphold and advance the honor, dignity, and status of the quality audit profession:

1. Each quality auditor will be honest and impartial, and serve with devotion employers, clients, and the public.

2. Each quality auditor will undertake only those audits compatible with the degree of training, experience, and proficiency he or she holds in regard to the technical or systems operations being audited.

3. Each quality auditor will demonstrate a freedom of mind and approach which will ensure objective viewing of the operation being audited.

4. Each quality auditor will be able to document the professional qualifications needed in order to provide clear, objective evidence of the degree of his or her technical and systems training.

6A.2 Relations with Employers and Clients

1. Each quality auditor will act in professional matters as a faithful agent or trustee of each employer or client.

2. The auditor will inform each client or employer of any business connections, financial interests, employment history, or affiliations which might influence, or appear to influence, his or her judgment or impair the equitable character of his or her services.

3. Independent auditing companies will maintain records of the auditors they employ showing any connections or associations of those auditors with their clients through previous employment or financial holdings.

4. Internal auditors will have their independence clearly defined by organizational policies and procedures.

5. The development of quality programs by auditors on a consultation

basis jeopardizes the independence of those individuals in their ability to assess the suitability of those programs in regard to meeting the requirements of standards. Auditors will always make such situations known to employers and prospective clients.

6. Auditors will issue reports which clearly define the degree of conformance or nonconformance of the operation being audited. In all cases the requirement against which conformance is being measured will be clearly defined.

7. Auditors will indicate to employers or clients the adverse consequences to be expected if their professional judgment is overruled.

8. No auditor will disclose information concerning the business affairs or technical processes of any present or future employer or client without obtaining consent to do so.

9. Auditing company records will clearly define the technical and organizational qualifications of each auditor. These records will include clear definitions of the areas and degree of expertise the auditor possesses that are applicable to the audit function.

6A.3   Relations with the Public

1. No quality auditor shall represent himself or herself to the public as a qualified auditor unless he or she can demonstrate the qualifications claimed.

2. Auditors will be dignified and honest in explaining their work and its merits.

3. Auditors will preface any public statement they may issue to clearly indicate on whose behalf it is being made.

6A.4   Relations with Peers

1. Each auditor will take care that the credit for all work done is given to those to whom it is due.

2. Each auditor will endeavor to aid the professional development and advancement of any employees or individuals under her or his supervision.

3. No auditor will compete unfairly with others, but each will extend friendship and confidence to all associates and business contacts.

4. Auditors will be encouraged to attend courses, seminars, conferences, etc., to broaden their knowledge of the various areas of the quality profession and of the quality audit in particular.

5. In audit projects where multiple auditors are involved, the team

leader or supervisor will be clearly identified. The leader will have demonstrated leadership during previous audits.

6. Each auditor will be expected to contribute to the development of improved techniques and methods within the quality audit profession and the organization that employs him or her.

7. The reports issued by auditors will be constructive in nature, pointing out improvements that can be made in those areas where shortcomings are noted.

8. The findings shall be quantified wherever possible.

Chapter

7

The Quality Auditor

7.1 Introduction

Successful quality auditors demonstrate an ability to work professionally and competently when exposed to a variety of situations. Many of these situations will be fraught with dichotomies resulting from real and apparently conflicting requirements. Examples of these dichotomies are given in Figure 7.1. These and other similar conflicting situations require that valid quality decisions be made, on a regular basis, by auditors. The ability to cope with these situations is one of the significant factors separating quality auditors from inspectors.

Therefore, it is important that quality auditors be carefully selected. The following paragraphs describe some of the professional and character traits that can help in the selection of candidates and the development of successful auditors. Obviously, such selection will involve compromise, as it is doubtful that any individual paragon, excelling in all of the points discussed, actually exists.

Of the standards reviewed by the author, only the American and Canadian quality audit standards—ANSI/ASQC Q1-1986 and CAN-CSA-Q395-1981—address the issue of requirements for auditors. As these two standards are being used as resource documents by the Quality Audit Working Group of ISO TC 176, hopefully the ISO standard when issued will reflect similar concerns. Both the Canadian and American standards, generic in nature, were prepared by writing committees representing broad cross sections of industrial and service disciplines. Therefore, the principles they espouse can apply to a very wide range of the procedures and techniques used by quality auditors.

7.2 General Aptitudes of the Auditor

The aptitudes and personality traits to be sought in a quality auditor can be readily subdivided into two general categories:

1. Situation:

The quality system should be planned and documented to ensure adequate control.

However,

The quality system must not inhibit creativity in management, design, methods, analysis, etc.

2. Situation:

The quality audit must be planned and systematic, using some form of working papers to ensure that all facets are covered.

However,

The quality audit must not be bureaucratic and only recognize one preconceived method of implementing the quality system. Each program and methodology must be assessed on its own merits.

3. Situation:

The quality audit compares actualities with the requirements of the reference documents.

However,

The reference documents will frequently have more than one interpretation and thus the quality assessment must be made on the basis of the intent of the document.

4. Situation:

The quality audit is a critique of the quality system or program.

However,

The quality audit must reinforce the system's strong points while identifying its shortcomings.

5. Situation:

The quality audit must provide objective evidence of compliance with the requirements in its report to the client.

However,

Management is frequently only concerned with control by exception.

6. Situation:

The quality system is adequately defined in policies and responsibilities and is working satisfactorily.

However,

The procedures or work instructions necessary to control and provide continuity in the activities have not been prepared.

7. Situation:

The output being audited meets all its requirements and there is supporting objective evidence to prove so.

However,

Verification is by means of process control or sequential inspection by the next operator after the one performing the operation and not by an inspector, independent of the supervisor of the productive function, as required by ISO 9001-1987 and ISO 9002-1987 (also a requirement of some other quality system standards).

Figure 7.1   Typical dichotomies facing the quality auditor.

1. Professional aptitudes

2. Other technical knowledge and personality traits or aptitudes

Actually, both of the above categories involve technical knowledge and personal traits or aptitudes. Those falling into the first directly affect the professionalism of the quality auditor, whereas those in the second are skills and abilities or attitudes that support those in the first.

A competent quality auditor will have a broad range of quality and industrial or service expertise, augmented by certain key personality traits. These traits are applicable to both areas of expertise. Figures 7.2a and 7.2b list typical areas of expertise and personality traits to be considered. The lists are indicative of the breadth of skills and knowledge desired but are not intended to include all the desired attributes; nor is the same level of expertise expected in all the areas of professional competence.

A quality audit is frequently conducted by a team of auditors. In this case, there must be a clearly identified team leader. The lead quality auditor must be able to demonstrate competence in all of the applicable areas to be covered by the audit. The other members of the

Factor or aptitude	Auditor	Lead auditor	Discussed in section
Qualifications			
Industrial or service discipline	x	x	7.3.1.1
Quality and reliability	x	x	7.3.1.2
Quality Auditing	x	x	7.3.1.3
Registered professional (chartered) status	x	x	7.3.2
Certification			
Quality Engineer	x	x	7.3.3
Reliability Engineer	x	x	
Quality Auditor	x	x	
Knowledge of standards—industrial, national, and international			7.3.4
Industry	x	x	
Service	x	x	
Product	x	x	
Process	x	x	
Quality systems	x	x	
Reliability systems	x	x	
Quality audits	x	x	
Knowledge of management systems and styles	x	x	7.3.5
Ethics	x	x	7.3.6
Integrity	x	x	7.3.7

Figure 7.2a Professional aptitudes needed by the quality auditor.

Factors or attributes	Auditor	Lead auditor	Discussed in section
Technical aptitudes			
Knowledge of technical standards—national and international			7.4.1
Industrial or service	x	x	
Quality engineering	x	x	
Reliability engineering	x	x	
Knowledge of cost accounting		x	7.4.2
Knowledge of quality cost systems	x	x	7.4.2
Knowledge of statistical techniques			7.4.3
Sampling	x	x	
Analysis	x	x	
Inferences	x	x	
Knowledge of diagnostic techniques			
Problem solving	x	x	7.4.4
Technical troubleshooting	x	x	
Personal attributes			
Personality traits			
Leadership		x	7.5.1
Interfacing abilities	x	x	
Confidence	x	x	
Composure	x	x	
Independence	x	x	7.5.2
Planning ability	x	x	7.5.3
Understanding of investigative techniques	x	x	7.5.4
Communications ability			
Oral	x	x	7.5.5
Written	x	x	
Report writing		x	
Critiquing ability	x	x	7.5.6
Decisiveness	x	x	7.5.7

Figure 7.2b Technical and personal traits and attributes needed by the quality auditor.

team may require specialized technical knowledge in specific areas, or a more shallow knowledge of and experience in the broader field of concern. An auditor operating alone should meet the requirements of a lead auditor.

7.3   Professional Aptitudes

The professional aptitudes needed by a quality auditor are a blend of technical knowledge and personality traits that allow him or her to carry out the audit activities competently and with integrity while creating in the auditee's mind a feeling of confidence and trust. Those traits are illustrated in Figure 7.2a. The auditor must be well versed in the expertise required by the applicable industrial or service discipline, as well as in the broader fields of quality and reliability engineering themselves. This does not imply that auditors must be able to perform or carry out every work function being audited, but rather that they must be able to understand the principles and objectives of each activity in order to provide an objective assessment of its suitability and effectiveness in regard to production and quality requirements.

The key to a professional approach is the ability to demonstrate an ethical and knowledgable approach to the audit. This gives the auditee a feeling of confidence in the auditor, the actual audit, and the resulting report. Certain of these skills and abilities can be taught in formal training programs. Others are inherent in the individual and developed through experience and maturity.

7.3.1 Qualifications

7.3.1.1 Industrial or service training   All auditors should have the academic training compatible with the needs of the industry or service they will be auditing, at a level expected of at least middle management. This can be either an implied or a mandatory requirement of one or more university degrees in the appropriate subjects. In those cases where a college degree is not required, the minimum level of education expected would be completion of a formal educational program leading to at least graduation from a secondary school (university entrance equivalent).

7.3.1.2 Quality and reliability training   Access to formal education in the areas of quality and reliability varies greatly from locality to locality. But with the interreactions that occur between the two disciplines, it is essential that all auditors have a thorough knowledge of both. They should have a knowledge of the two areas that at least meets the level required for certification by The American Society for Quality Control as Certified Quality Engineers (CQE) and Certified Reliability Engineers (CRE) (see Section 7.3.3). The knowledge may have been obtained from formal training courses, special courses, seminars, correspondence courses, a planned reading program, etc. If the qualifications have been obtained outside a course of formal education, some kind of certification, such as that offered through the ASQC programs, becomes important as a demonstration of capability.

7.3.1.3 Quality auditor   Once a general knowledge of the quality and reliability fields has been obtained, an expertise in auditing must be developed. Training course, seminars, etc., are available for this purpose. Again, an ASQC certification program is one of the best ways of demonstrating this knowledge. Other auditor registration programs are available in specialized fields or areas of jurisdiction.

7.3.2 Professional or chartered status

In some countries, states, provinces, etc., there are regulations that require an individual carrying out certain quality audit activities to be qualified and registered as either "professional" or "chartered" in the associated field of expertise, e.g., engineering, accounting, medical, etc. This requirement can apply to either external or internal quality auditors. Normally, it only applies where an auditor is certifying products or services, or systems controlling products or services, that have a direct impact on the safety and well being of the general public.

7.3.3 Certification—quality and reliability engineering and quality auditing

An effective way of demonstrating knowledge in the fields of quality and reliability engineering and quality auditing is by obtaining certification in these categories from The American Society for Quality Control. ASQC's certificates in the three fields have achieved international recognition as measures of excellence on the part of those holding them. The certificates are not restricted to members of the society and so may be obtained by anyone qualified as evidence of knowledge and capability. Several other associations, societies, agencies, etc., also issue certificates in these fields showing the level of expertise obtained by those holding them.

7.3.4 Standards—industrial, national, and international

As trade in both industrial products and services crosses more and more national boundaries, the need for an awareness of national and international standards becomes more and more important for the quality professional in general, and the quality auditor in particular. Most industries or services have their own family of standards issued as either industrial or national standards for use within a given country. There are also numerous international bodies involved in developing product and service standards, with two of the most active being headquartered in Geneva—namely, the International Organization for Standardization (ISO) and the International Electro-Technical Committee (IEC).

In the quality and reliability fields, the national standards-issuing agencies listed in Appendix 1A.7 form a good cross section. Each of them publishes an index of standards, which can be obtained from the addresses shown in the appendix for the agencies; a fee may be charged. In the international arena, ISO is responsible for developing generic quality standards (TC176) and statistics (TC69), while IEC concentrates on reliability standards (TC56), initially for the electrical and electronic industries alone but now with a broader scope.

7.3.5 Management systems and styles

The quality auditor should have a basic understanding of the various types of management systems and styles that exist. Many of them are discussed in various business textbooks, business school classes, etc. As auditors gain experience through auditing different organizations, they will gradually be exposed to the practical applications of most of the traditional types, e.g., Management by Objectives (MBO), Theories X and Y, Theory Z, participatory management, dictatorial management, etc. Regardless of an auditor's past experience with any particular type or the management style preferred by him or her, each new audit must be approached objectively no matter what the management system or style used by the organization being audited.

7.3.6 Ethics

A formal code of ethics is required for quality auditors to give a uniformity of approach to this important topic. Currently, no such code has been approved by any standards body, although one has been submitted by the author to the Quality Audit Technical Committee of The American Society for Quality Control (see Appendix 6A).

A formal code would provide a benchmark against which the auditee and the client could measure the auditor's activities, independence, and potential conflicts of interest. There are too many facets to this topic that can easily be overlooked without a formal code recognized by all parties. Moreover, the lack of such a code leaves the ethics of auditors and auditing organizations subject to measurement and comment against some undefined and varying reference. This does not provide a sound foundation for the development of and judgment about reputations.

Personally, I believe that each external auditor should provide a signed copy of a code of ethics to the client and the auditee at the initiation of each external quality audit. This would give both parties a feeling of confidence in the knowledge, honesty, and ethics of the auditor. And a signed copy of such a code should form part of the personnel record of every internal auditor, with copies also being provided to members of the organization's management team to ensure that the expected performance of the auditor is known by all concerned.

7.3.7 Integrity

Whether operating as an internal or an external quality auditor, a prime requisite for both an individual auditor and an auditing organization is a reputation for integrity. This can only result from demonstrated honesty and forthrightness in conducting and reporting assignments, as well as demonstrated trustworthiness in protecting the proprietary rights and confidentiality of data relating to auditees.

The success of a quality audit depends upon a free and open exchange of information between the auditor and the auditee. If the auditor has lost a reputation for integrity, this exchange will not take place. This loss could even lead to an adversarial environment, which virtually destroys the effectiveness of many audits.

Every auditor must be cognizant of the proprietary rights of the auditee. Reports resulting from an audit must not infringe these rights. If products, processes, or services involving proprietary rights are mentioned in the report, the comments should be carefully couched in terms that will not infringe these rights. Alternatively, where more open comments are necessary, a controlled distribution of the report can provide the necessary protection. Similar precautions are essential for any off-the-record discussions.

An internal quality auditor must work for an organization having corporate integrity if the auditor is to earn and retain a reputation for personal integrity. An organization with integrity will provide the necessary positive reinforcements to the auditor to encourage and support the development of his or her personal integrity. For example, such an organization is unlikely to abuse the data derived during audits by taking punitive action against individuals. The absence of integrity in an organization can easily lead to reprisals against an auditor who has integrity. In some extreme cases, internal quality audits requesting corrective action have been classed as a form of whistle blowing, with all of that term's connotations. This indicates a lack of understanding of the principles and objectives of a quality audit.

7.4 Technical Attributes

The technical attributes required by quality auditors are those that form the body of knowledge related both to the quality and reliability professions and the various industrial or service disciplines they will be auditing. Figure 7.2b illustrates some of the areas concerned. This listing is intended to serve as a memory prompter for an auditing organization developing the profile of traits it desires in its auditors.

7.4.1 Technical standards and regulations—industrial, national, and international

Technical standards, codes, and regulations fall into four major categories:

1. Those defining requirements applicable to the product of the organization, e.g., atomic energy standards, regulations, codes, etc., boiler codes, aviation safety regulations, electrical codes and standards, telephone interface standards, etc.

2. Those defining the various productive processes used within an industrial or service discipline, e.g., welding, soldering, heat treatment, curing, etc.

3. Those defining the services, goods, and materials used in preparing the ultimate product, e.g., ASTM materials standards, military hardware and component standards, software standards, etc.

4. Those involved with the control of a process or its verification, e.g., process control standards, metrology standards, statistical standards, production control standards, etc.

Standards, codes, and regulations in each of these areas are issued by related industrial or professional associations, by national standards writing organizations concerned with the intended marketplace, and in many cases by international bodies such as ISO, IEC, NATO, the International Atomic Energy Agency (IAEA), etc. The bringing together of the various standards and regulations covering a particular discipline can be a major exercise. Becoming familiar with the requirements of a field is a major task that needs regular updating.

7.4.2 Cost accounting and quality cost systems

A basic knowledge of cost accounting is of value to a lead auditor or a lone operator, as it will frequently be necessary to review with the accounting department of an auditee variance costs involved with particular observations as one of the means of assessing the criticality of points in question.

When auditing companies that use some form of quality cost system, a knowledge of cost accounting will be essential in evaluating the effectiveness of that element of the overall quality system, especially a knowledge of its principles and objectives. As there will be many ways in which quality cost figures are developed, it is essential that the auditor not take a dogmatic approach to evaluating these systems.

7.4.3 Statistical techniques

As with so many areas of the quality and reliability professions, the quality auditor must have a thorough knowledge of certain kinds of statistical techniques, including:

1. *Sampling procedures.* All audits depend on data derived from samples. It is virtually impossible for an auditor to evaluate any activity on the basis of 100 percent examination and evaluation. Therefore, it is essential that *all sampling plans be based on sound statistical principles*. Wherever possible, use should be made of recognized sampling plans, known by both the auditor and the auditee. In this way, both parties will be aware of the risks involved. (See Chapter 10).

2. *Interpretive techniques.*   Wherever possible, a quality audit should be reported on a quantitative basis. In this way, trends can be determined over a series of audits. Quantified results also provide a measurable benchmark for determining the effectiveness of any changes introduced.

3. *Confidence levels.*   Confidence levels will normally be used in conjunction with the interpretive techniques. However, their special importance arises through their use in reports as a means of indicating the confidence that can be placed in the quantified data. An auditor must be able to use these factors without confusing the reader of the report. Hence, the wording can be crucial.

7.4.4 Diagnostic techniques

Diagnostic techniques involve two areas of expertise for those auditors involved with products as well as processes, namely:

1. *Problem solving.*   Finding solutions to observed problems
2. *Troubleshooting.*   Finding reasons for equipment malfunction

"Problem solving" is an organized, systematic means of discovering the actual cause of a faulty process, as distinct from determining the various symptoms revealing the fault. "Process" in this sense covers any activity where a number of things must happen to produce an end result.

There are many forms of problem solving covered in a variety of texts and training courses. The auditor should be knowledgable and skilled in at least one such technique and be aware of others in order to be able to:

1. Determine the cause of any problem noted during the audit

2. Evaluate this element of the quality program

Most problem-solving techniques follow the same fundamental logical sequence of steps:

1. Determining the various symptoms showing that a problem is present.

2. Examining each symptom to determine what can and cannot initiate or cause that symptom directly or indirectly.

3. Analyzing the causes identified above to place priorities on them, ranging from most likely to least likely. Pareto distributions can be of value in more complex situations.

4. Investigating each cause, in turn, to determine which is present and causing the current problem.

5. Determining and implementing the corrective action necessary to eliminate the cause.

6. Determining the effectiveness of the corrective action.

"Troubleshooting" is the application of logical testing sequences to equipment in order to determine the cause of a malfunction. The various techniques are designed to zero in on the faulty part or component with a minimum number of tests and thus in the shortest time possible. An auditor should be knowledgable in these techniques if his or her auditing work involves evaluating the functions of products or equipment.

7.5   Personal Attributes

7.5.1   Personality traits

To a large extent, the effectiveness of a quality audit depends on the personal relationships between the auditor and the members of the auditee's staff. The auditee's personnel must have confidence in the objectivity, competence, expertise, and ability of the auditor to make valid judgments.

Preselection personality testing can identify some of the traits necessary for a successful quality auditor, as well as identify traits that may be harmful in an audit situation. Identifying some of these will require special testing methods to determine what the individual's response would be under the pressures of an actual audit.

The key personality traits necessary for a successful quality auditor include the abilities to:

1. Deal with people in a manner which will inspire a free and open exchange of facts and ideas
2. Remain cool and calm throughout all the stages of a quality audit
3. Exude an air of confidence, but not overconfidence or cockiness
4. Display honesty and forthrightness in reporting the findings of an audit
5. Demonstrate professional knowledge of the quality field and appropriate industrial or service disciplines to the client, auditee, and their associates, and to the remainder of the audit team
6. Work in a planned and systematic manner, without showing autocratic tendencies
7. Be decisive in resolving questions about the suitability of, and the conformity to, a quality program, without being bureaucratic or dogmatic
8. Be broad-minded within the terms of the reference documents, standards, etc.
9. Be observant of details in associated areas and activities
10. Be intuitive in interpreting the observations made during an audit, in order to identify and react to any anomalies that may arise
11. Provide leadership, if being considered for the position of lead auditor

7.5.2 Independence

A quality auditor must have an independence of mind and spirit to be successful in either internal or external quality audits. This independence must also be apparent to others. Thus the auditor must not only be, but must also be clearly seen to be, independent.

However, there are differences in the degree of independence that various internal and external auditors have. These arise through an auditor's participation in the development of and/or working as part of a quality system. In either of these situations, an auditor, whether internal or external, has jeopardized his or her independence and objectivity in relation to assessing the suitability of that quality program with respect to a particular quality system standard (although not in relation to undertaking conformity quality audits).

In addition, an external quality auditor should not in the past have been employed by, currently have financial holdings in, or in other ways maintain business connections with the organization being audited. The auditor must not be beholden to the auditee in any way. Similarly, as a protection to the auditee, the auditor should not be beholden to any major competitor of the auditee, (although relations to competitors do not affect an auditor's independence in undertaking vendor appraisal or vendor surveillance types of audits for clients). Ideally the auditor should have a personal resumé that demonstrates this independence. This could imply preparing customized resumés for different clients and auditees.

The employment of a quality auditor by a customer or a potential customer should not disqualify that auditor from conducting either a suitability or a conformity quality audit on a supplier, providing the auditor has not been involved in the development or application of the supplier's quality system.

7.5.3 Planning ability

All quality audits will be disturbing to one degree or another to the operations of the auditee's facility. Therefore, it is essential that the audit be planned to minimize these disturbances as far as possible. The details of planning needed are discussed in Chapters 8 and 11, "System Analysis" and "Implementation of the Quality Audit," respectively. Suffice it at this point to indicate that auditors must be trained in various systems analysis tools and be knowledgable as to their advantages and disadvantages. This will enable them to select the proper tools for the job at hand.

Planning must be thorough, but must have the flexibility necessary to help participants cope with unexpected incidents when they occur during an audit.

Planning should also be formal and documented, so that all parties are aware of the intent or objective of the audit and the general modus operandi by which it will be conducted.

7.5.4 Ability to use investigative approach

A quality audit requires a thorough knowledge of and the patience to use investigative techniques, in order to ensure that all the pertinent facts are being determined. An auditor must be able to phrase questions in a manner that will be understood by the auditee's representatives, without speaking down to or over the head of the individual being questioned. The auditor must also be aware of when to explore a topic in greater depth and when to drop it.

7.5.5 Ability to communicate

A successful quality audit depends on two-way communication, with the auditor transmitting requests for information mostly verbally and occasionally in writing, but receiving information through all the senses—hearing, sight, smell, feel or touch, and taste. The flow of information to the auditor through the senses is continuous throughout the audit, and the auditor must be aware of it.

The most error-prone sense is hearing, normally not because of any lack of sensitivity in the ear but because of the inability of many individuals to listen properly—that is, the combination of mind and ear misinterprets what is being heard. Since verbal communications are the primary mode of information flow during an actual audit, the need for care in listening cannot be stressed too strongly.

Verbal data will come from individuals in both technical and nontechnical fields that have varying degrees of linguistic fluency. It will frequently include technical jargon or terms having special meanings within a specific discipline, organization, etc. Therefore, it is important that the auditor be able to converse proficiently with the persons providing the data. This implies that the auditor must be able to adjust his or her terminology and phrasing to match the linguistic ability of the individual being questioned. Therefore, it is important that the auditor be fluent in both the technical and the nontechnical languages of the client and auditee. This need for fluency extends to all written communications, and is particularly important in the audit report which is the *raison d'être* of the audit.

To improve the quality of verbal communications during an audit, the following actions by the auditor are recommended:

1. The auditor should carefully and clearly transmit all requests for information.
2. The auditor should carefully *listen* to the reply.
3. The information transmitted back should then be retransmitted back to the auditee's representative to show how the reply has been interpreted.
4. If the auditee does not agree with the retransmitted answer, the cycle should be repeated after clarifying the original request for information.

This approach should result in a common understanding of the question and the reply, thereby increasing the accuracy of the audit.

Besides hearing, great care should also be taken in interpreting the signals received by the other senses—sight, smell, touch, and taste. That information must be confirmed through questioning. Generally speaking, the signals detected by these four senses will be an indication of changes in atmosphere or surroundings and thus are qualitative in nature.

Similarly, care must be taken with all written communications, as the messages they convey are transmitted only by words with no tonal intonations or body language to help interpret their statements. The auditor must be sensitive to the nuances of written language and choose appropriate wording. All written communications should be couched in terms appropriate to the fluency of their recipients. Augmenting the text with appropriate tables, graphs, and other illustrations will assist in communications.

7.5.6 Understanding nature of critiques

A quality audit is a critique of a quality system, and thus the auditor should be familiar with the various techniques for conducting critiques. It is essential that the auditor recognize a critique's results can be both positive and negative. A critique is a constructive criticism that allows the auditee to become aware of the strengths and weaknesses in the quality system being audited. With the present-day emphasis on management by exception as a means of efficient use of time, this point is frequently overlooked. It is important to remember that a quality audit must *not* be a destructive force, solely accenting the errors or shortcomings in the quality system.

7.5.7 Decisiveness

A quality auditor must be able to make sound and valid quality decisions. All the questions that arise about the suitability of a quality program and its conformity to quality standards must be decisively resolved. Except for auditors in training, it is the auditors conducting the audit who are responsible for making those decisions and defending them if necessary. In cases where a team is conducting the audit, the lead auditor should *not* make decisions on behalf of subordinate auditors. However, the lead auditor should review the decisions made by those subordinates.

It is extremely important that auditors be thoroughly familiar with the principles of "decision making." In addition to developing a decisive approach, auditors must be able to evaluate decisions as an element of decision sampling. The following logical sequence is normally used in effective "decision making":

1. Clearly identifying the nature of the decision to be made, i.e., separating the symptoms from the underlying cause, as in problem solving.

2. Carefully determining what the options or alternative decisions

are. *Note:* Decisions will seldom be simple binomial situations of A or B.

3. Weighing the impact of the various alternatives, including both positive and negative factors.

4. Selecting the most effective decision, i.e., the one showing the highest positive value.

5. Once a decision has been made, evaluating its results.

6. If the decision is shown to be effective, letting it stand. However, if it is shown to be ineffective, reexamining the options—old and new, their positive and negative points—and being prepared to make a new decision based on this latest data.

7. Measuring the effectiveness of the new decision.

Being decisive does not imply being dogmatic about a decision that has been shown to have shortcomings, nor does it imply snap decisions that do not consider all the known facts. If a decision is later found wanting, it is best to admit a better decision could have been made and work toward that new and better decision.

There are many reference texts on decision making. Copies of selected works should be in the library of any auditing organization. One of the most effective is *Decision Making* by Irving L. Janis and Leon Mann (see the Bibliography at the back of this book for publication details).

7.6 Conclusion

The above paragraphs outline some of the skills and attributes we all would like each quality auditor to have. This paragon is a rather rare and distinguished individual. Thus in most practical cases, a quality auditor will be selected on the basis of a compromised rating of the various factors discussed. One key type of individual that should be avoided at all costs is the one who looks on auditing as a source of power over other members of the organization. In auditing we are looking for individuals with integrity who are thorough, think creatively, and are able to work with people.

System Analysis

8.1 Introduction

System analysis from the quality audit point of view falls into two general categories:

1. Analysis of the reference documents or standards
2. Analysis of the auditee's system at the level at which it is to be evaluated

In both cases, the analysis is used to identify each element of the function concerned, to ensure that each element is addressed during the subsequent audit.

The extent of the analysis of an auditee's system will depend on the purpose of the audit. Where a quality audit is being carried out to determine the degree of compliance with a particular reference quality standard, only those activities relating to the standard should be the subject of the analysis, audit, and subsequent report. In other cases, where an audit is intended to evaluate the effectiveness of an overall quality system, the analysis must cover all the functions in the auditee's organization that impinge in any way on that quality system.

There will be two major outputs of this analysis:

1. Working papers, i.e., audit instructions, memory prompters, checklists, etc., covering each element to be audited
2. An overall audit plan collating all applicable detailed working papers into the desired operating sequence

Techniques or tools for systems analysis are discussed in many texts and papers, in a variety of fields and disciplines. They have originated in various fields such as accounting, operational research, manufacturing planning, information systems, quality systems, etc. Most of these will be valuable, to varying degrees, in planning a quality audit.

This chapter will briefly cover a few of the techniques used by the author in developing quality audits for various activities. These include:

1. The function tree (discussed in Section 8.2)

2. The decision flow chart (discussed in Section 8.3)

3. The critical path network (discussed in Section 8.4)

4. The matrix responsibility chart (discussed in Section 8.5)

5. Function tables (discussed in Section 8.6)

6. Ishikawa (cause and effect) diagram (discussed in Section 8.7)

8.2   Function Tree

The function tree is a means of ensuring that each function of an organization being audited is broken down into its most fundamental elements. This is normally done diagrammatically, as shown in Figure 4.1, with the major functions on the left and the subfunctions in columnar presentation to the right. Properly applied, this is the strongest technique available for ensuring that each element of an activity is identified and addressed.

This technique has been discussed briefly in Sections 4.4, 4.5, and 4.6. It has also been used as a means of analyzing the functions of the quality audit itself.

When developing a function tree for a reference standard, the primary function will be the purpose or objective of the standard. The first tier or level of subfunctions in the tree will be the major functional headings of the standard. Subsequent levels will be the various functions referred to in the text, with any subparagraph headings taking priority over the text references contained in the subparagraphs.

The majority of procurement quality standards reviewed by the author in preparing this text define elements comprising a quality system and tend to ignore the overall objectives of those elements. This concentration on techniques has a tendency to inhibit the development of new or improved systems. To avoid this, it is the author's opinion that procurement quality standards should define the desired functions. For example, a standard could note the following:

1. The primary function is to provide assurance that the items delivered against the contract meet the needs of that contract.

2. The secondary level consists of the major functions needed to achieve

that aim, which could include market research, design, material, production, sales, etc.

The standard should also indicate the degree of assurance required to confirm that each particular operation has been carried out to the satisfaction of the customer.

Where these functional requirements are omitted in a standard, the author recommends they be inserted into the function tree at the appropriate levels. In this way the purposes of the activities will not be lost.

Typical function trees are shown in Figures 8.1 and 8.2 for CAN-CSA-Z299.1-1985 and ISO 9001-1987, respectively.

When developing a function tree, such as the one shown in Figure 8.3, for an organization desiring a improvement in its quality program, it is essential that the major functions analyzed fall within the scope of the quality audit. Thus for a design and manufacturing organization, the analysis might address all or some of the following typical functions having an impact on the organization's quality system:

1. Marketing or contract administration

2. Design

3. Materials control, including procurement

4. Production

5. Packing and shipping

6. Customer service

7. Installation and commissioning

8. Disposition after use

9. Accounting

10. Quality organization, of whatever name and size

Similar functions, with some variations in title, exist in most service organizations.

Once the systems function tree has been developed, analysis commences from right to left. A working paper is required for each different element or subfunction identified. Often these papers will have multiple applications, since the same subfunctions may exist in different areas of the organization; the function tree will identify these instances. When all of the papers have been completed and referenced to the function tree, the tree becomes a systems control document for the actual audit to ensure that no activities are overlooked. Thus a collated set of working papers can be developed to serve as the master control document for the audit, while copies of particular pages are

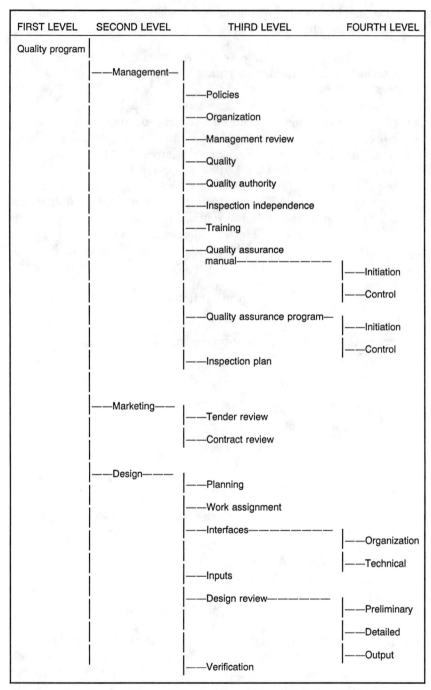

FIRST LEVEL	SECOND LEVEL	THIRD LEVEL	FOURTH LEVEL
Quality program			
	——Management—		
		——Policies	
		——Organization	
		——Management review	
		——Quality	
		——Quality authority	
		——Inspection independence	
		——Training	
		——Quality assurance manual———————	——Initiation
			——Control
		——Quality assurance program—	——Initiation
			——Control
		——Inspection plan	
	——Marketing——		
		——Tender review	
		——Contract review	
	——Design———		
		——Planning	
		——Work assignment	
		——Interfaces———————	——Organization
			——Technical
		——Inputs	
		——Design review—————	——Preliminary
			——Detailed
			——Output
		——Verification	

Figure 8.1 Function tree for a quality assurance program based on CAN-CSA-Z299.1-1985, Category 1.

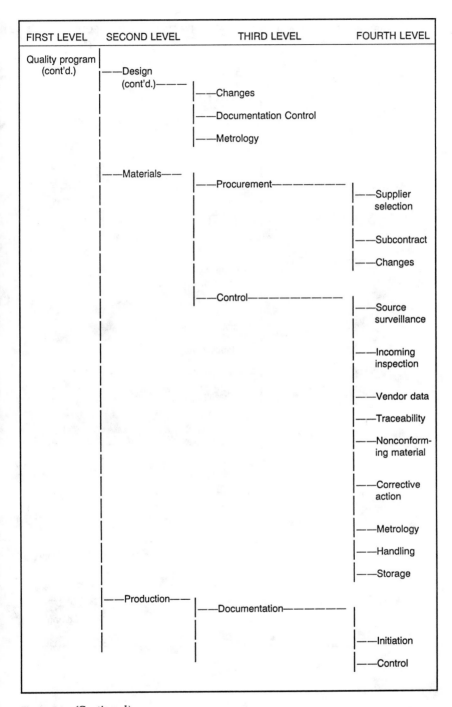

Figure 8.1 *(Continued)*

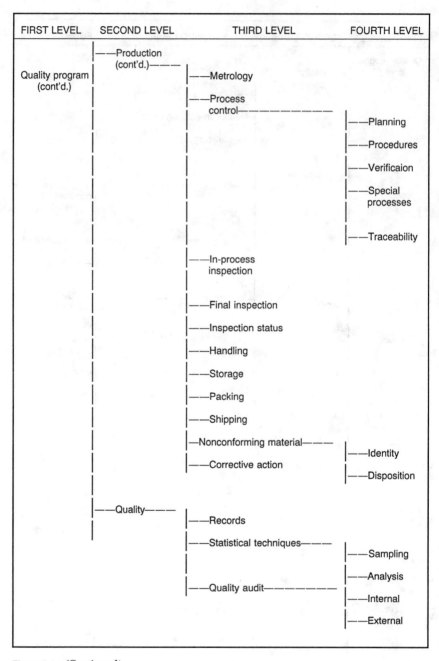

FIRST LEVEL	SECOND LEVEL	THIRD LEVEL	FOURTH LEVEL
Quality program (cont'd.)	——Production (cont'd.)———	——Metrology	
		——Process control———————————	——Planning
			——Procedures
			——Verificaion
			——Special processes
			——Traceability
		——In-process inspection	
		——Final inspection	
		——Inspection status	
		——Handling	
		——Storage	
		——Packing	
		——Shipping	
		—Nonconforming material———	——Identity
		——Corrective action	——Disposition
	——Quality———	——Records	
		——Statistical techniques———	——Sampling
			——Analysis
		——Quality audit——————————	——Internal
			——External

Figure 8.1 *(Continued)*

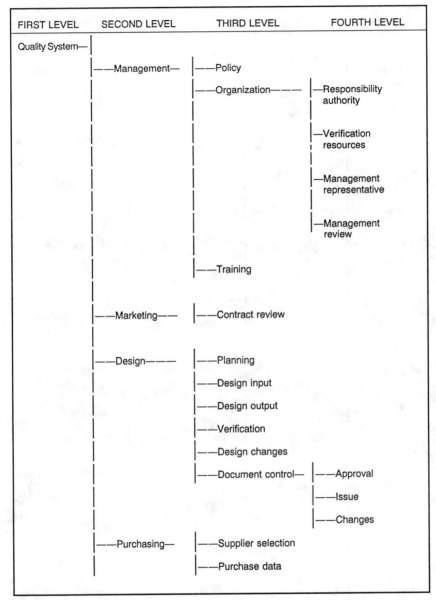

FIRST LEVEL	SECOND LEVEL	THIRD LEVEL	FOURTH LEVEL
Quality System—			
	——Management—	——Policy	
		——Organization———	—Responsibility authority
			—Verification resources
			—Management representative
			—Management review
		——Training	
	——Marketing——	——Contract review	
	——Design———	——Planning	
		——Design input	
		——Design output	
		——Verification	
		——Design changes	
		——Document control—	——Approval
			——Issue
			——Changes
	——Purchasing—	——Supplier selection	
		——Purchase data	

Figure 8.2 Function tree of a quality systems model for quality assurance in design/development, production, installation, and servicing, based on ISO 9001-1987.

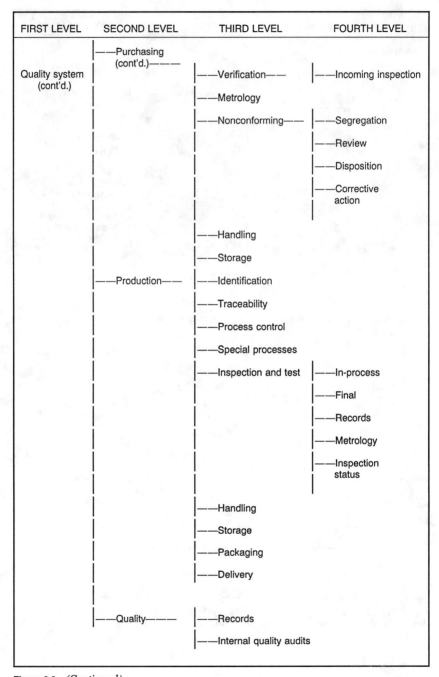

FIRST LEVEL	SECOND LEVEL	THIRD LEVEL	FOURTH LEVEL

Quality system (cont'd.) — Purchasing (cont'd.) —
- Verification — Incoming inspection
- Metrology
- Nonconforming — Segregation
  - Review
  - Disposition
  - Corrective action

— Production —
- Handling
- Storage
- Identification
- Traceability
- Process control
- Special processes
- Inspection and test — In-process
  - Final
  - Records
  - Metrology
  - Inspection status
- Handling
- Storage
- Packaging
- Delivery

— Quality —
- Records
- Internal quality audits

Figure 8.2 *(Continued)*

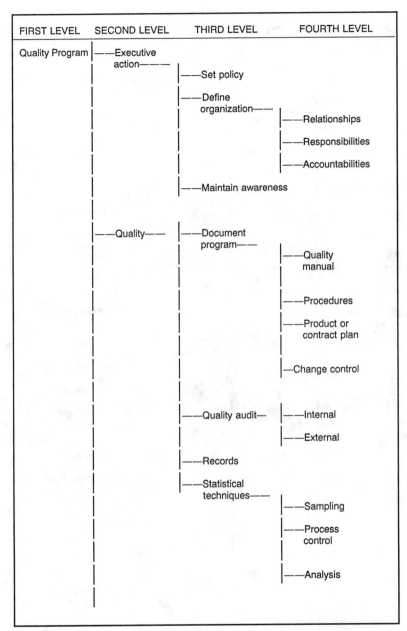

FIRST LEVEL	SECOND LEVEL	THIRD LEVEL	FOURTH LEVEL

Quality Program ——Executive action———
——Set policy
——Define organization——
——Relationships
——Responsibilities
——Accountabilities
——Maintain awareness
——Quality——
——Document program——
——Quality manual
——Procedures
——Product or contract plan
—Change control
——Quality audit—
——Internal
——External
——Records
——Statistical techniques——
——Sampling
——Process control
——Analysis

Figure 8.3 Function tree for a quality improvement program.

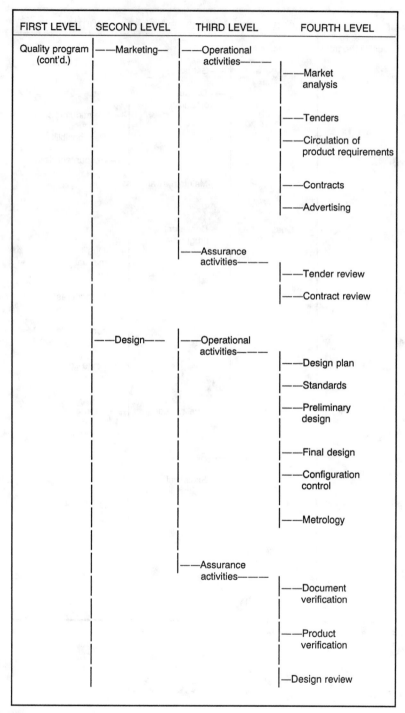

FIRST LEVEL	SECOND LEVEL	THIRD LEVEL	FOURTH LEVEL
Quality program (cont'd.)	——Marketing—	——Operational activities———	
			——Market analysis
			——Tenders
			——Circulation of product requirements
			——Contracts
			——Advertising
		——Assurance activities———	
			——Tender review
			——Contract review
	——Design——	——Operational activities———	
			——Design plan
			——Standards
			——Preliminary design
			——Final design
			——Configuration control
			——Metrology
		——Assurance activities———	
			——Document verification
			——Product verification
			—Design review

Figure 8.3 *(Continued)*

FIRST LEVEL	SECOND LEVEL	THIRD LEVEL	FOURTH LEVEL
Quality program (cont'd.)———	——Materials control——	——Procurement (operational)—	——Supplier selection
			——Vendor appraisal
			——Subcontract requirements
			—Change control
			——Expediting
		——Receiving———	——Documentation
			——Handling
		——Storage (operational)—	——Traceability
			——Protection
			——Shelf life
			——Handling
			——Withdrawal
			——Inventory control
		——Assurance——	——Quality terms and conditions for suppliers
			——Source surveillance
			——Vendor data
			——Vendor approval

Figure 8.3 *(Continued)*

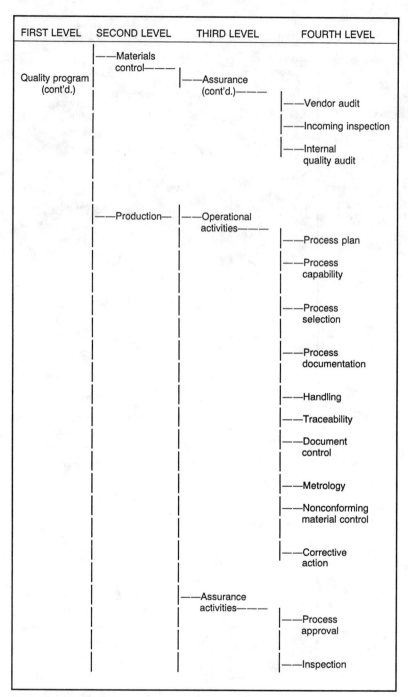

FIRST LEVEL	SECOND LEVEL	THIRD LEVEL	FOURTH LEVEL

Quality program (cont'd.)

——Materials control———

——Assurance (cont'd.)———

——Vendor audit

——Incoming inspection

——Internal quality audit

——Production—

——Operational activities———

——Process plan

——Process capability

——Process selection

——Process documentation

——Handling

——Traceability

——Document control

——Metrology

——Nonconforming material control

——Corrective action

——Assurance activities———

——Process approval

——Inspection

Figure 8.3 *(Continued)*

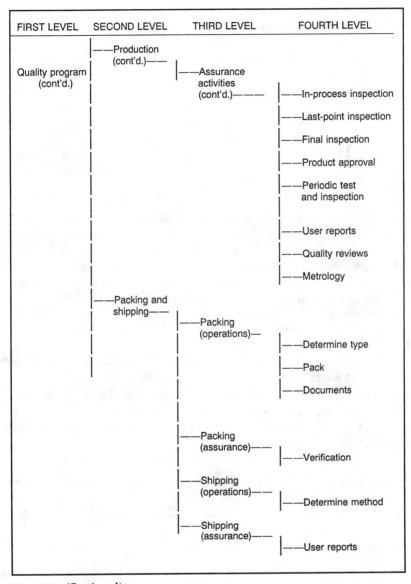

FIRST LEVEL   SECOND LEVEL   THIRD LEVEL   FOURTH LEVEL

Quality program (cont'd.)
——Production (cont'd.)——
——Assurance activities (cont'd.)———
——In-process inspection
——Last-point inspection
——Final inspection
——Product approval
——Periodic test and inspection
——User reports
——Quality reviews
——Metrology

——Packing and shipping——
——Packing (operations)—
——Determine type
——Pack
——Documents

——Packing (assurance)——
——Verification

——Shipping (operations)——
——Determine method

——Shipping (assurance)——
——User reports

Figure 8.3   *(Continued)*

used by the auditors designated to deal with the particular functions they show.

This analysis will also determine where to use location-oriented audits as opposed to function-oriented audits. Location-oriented audits are particularly useful where a number of subfunctions apply at several different locations. By using a location-oriented approach, any differences in the interpretations of those functions will be evident. For example, in a manufacturing department, working papers covering drawing control, metrology calibration, process control, control charts, materials identification, identification of nonconforming material, materials segregation, etc., could be applied to each cost center or location audited. Function-oriented audits are more applicable in activities such as materials control, calibration systems and facilities, etc. But note that the effectiveness of the calibration control system can be determined from findings taken from location-oriented audits.

8.3 Decision Flow Chart

A decision flow chart shows the sequence of events that occur in the particular process under evaluation. It depicts the various activities or actions through the use of symbols. There are various kinds of symbols to choose from, and stencil sets for each variety are available. These have been developed for systems analysts, industrial engineers, data process programmers, etc. Any of them can be readily adapted for use in analyzing a quality system. Because of the additional symbols it provides, I favor the data processing stencil. A typical set is shown in Figure 8.4.

The flow chart covers each activity that affects the end product or its component parts during the life cycle concerned. Similarly, for service activities, it covers the design, development, and implementation of the service concerned. Activities are entered sequentially, i.e., in the order in which they occur. If they happen simultaneously, parallel paths are introduced.

Fundamentally, there are four key symbols:

1. *Process symbol.*   This is used for each activity as it occurs. Activities include such functions as design, make, enter, assemble, drill, inspect, test, etc. Thus, the symbol covers all functions whether operational or assurance-type activities. Certain functions, such as inspection or test, are frequently followed by decision and documentation activities. Therefore, symbols for these would follow the process symbol identifying the initial activity.

2. *Document symbol.*   This signifies a recording activity on a new or existing form. Normally the form name or number is included. If multiple copies of the document are produced, multiple images of the symbol are frequently used. Each can then be separated to show the applicable "process" blocks.

3. *Decision symbol.*   This is used at any point where a decision is required as part of the flow. Typically a "decision" point follows each

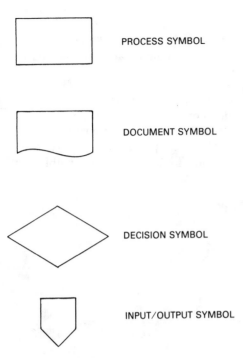

PROCESS SYMBOL

DOCUMENT SYMBOL

DECISION SYMBOL

INPUT/OUTPUT SYMBOL

Figure 8.4 Typical decision flow chart symbols.

"inspection" or "test" point. The symbol is designed for a binary type of decision, i.e., yes/no, OK/NOK, good/bad, etc. However in the interest of clarity, it is recommended that the question be phrased so that it can be answered only by yes or no. In this way unusual acronyms that may not be universally understood can be avoided. For example, an inspection process block might be followed by a decision symbol containing the question "Does material meet requirements?" This can obviously be answered by yes or no.

4. *Input/output symbol.* This is used to identify inputs from outside the flow chart and output(s) from the flow chart. These may lead to or from another flow chart in cases where multiple charts have been prepared in the interest of simplicity. Where this occurs, the appropriate chart identification should be given with the output reference.

In preparing a decision flow chart, some means must be included to show the departments or activities responsible for the various blocks. Typical ways include:

1. Entering the appropriate department number in the corner of the block.

2. Color-coding the blocks.

3. Entering marginal notations for departments, and using dotted lines to separate the departments. The symbols are then placed in the appropriate line or column.

A simplified decision flow chart is shown in Figure 8.5.

The decision flow chart technique is extremely effective in showing the sequence of events in activities such as flow of materials, paperwork, processes, etc. It can readily show the points where particular decisions or dispositions must be made and who is responsible for

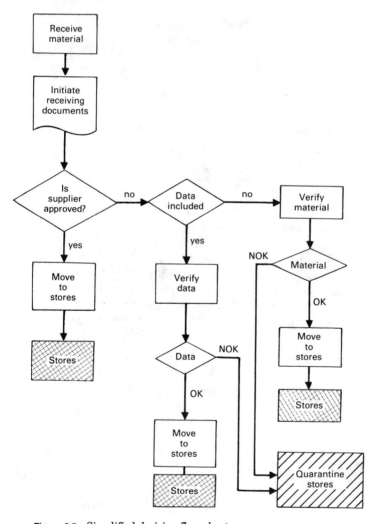

Figure 8.5 Simplified decision flow chart.

making them. It also clearly shows where activities occur simultaneously or in parallel through the use of branching or of separate charts drawn with the proper input/output symbols. These strengths make it a very powerful production planning guide or product inspection plan, as required by several procurement standards.

The technique has the disadvantage of not providing clear visibility for the procedures that apply to operations, or clear guidance to help identify those operations where the procedure itself is not a part of the sequence of events. This difficulty applies to procedures covering standards, guidelines, acceptance criteria, etc. To some degree such procedures can be shown as footnotes to the chart. However, in many cases the procedures would be too complex for the footnotes to be of value.

8.4 Critical Path Network

The critical path network (CPN), also frequently called PERT (Project Evaluation and Review Technique), is a production control or scheduling tool that can readily be adapted to analyzing the flow of goods and services from a quality audit point of view. It is frequently used to trace the activity flow, step by step, for the first-off unit of a production program, where successive units will be following the same sequence of events except for some special activities, e.g., verification tests, that are applicable to the first-off unit only. Thus CPN involves a chart showing each significant activity, including the various inspection, test, and other verification points. It provides, in effect, a detailed road map of the steps taken to convert an idea, material, services, etc., into a final product or service delivered to a customer.

A CPN chart is constructed according to the same general principles as a decision flow chart, but without the use of different symbols to distinguish between activities. CPN uses numbered points to indicate the start and completion of each activity. These numbers are normally defined on an accompanying table. Parallel paths are used to indicate parallel operations.

Since CPN was developed as a schedule control tool, the horizontal axis of the chart is normally scaled in units of time—elapsed days, weeks, dates, etc. The distance between the event symbols, i.e., between the start and completion symbols, represents the weighted average time for an operation.

Computer programs have been developed that produce CPN charts and then evaluate the times for the various activities to determine which operations govern the total elapsed time. Initial weighted estimates for the duration of each operation are based on the formula:

$$\text{Plotted time} = \frac{\text{minimum time} + 4 \text{ (expected average time)} + \text{maximum time}}{6}$$

The plotted times for each activity are then collated to determine which path connecting the various activities shows the longest period of time. This becomes known as the "critical path." The computer program normally produces a printout showing all the paths but highlighting the critical path in some way. Normally, these charts are updated at regular intervals to show the effect of changes caused by more accurate time, delay, etc. Each updating should result in a reevaluation of the data and could result in a new critical path. As the network is a mathematical model for a given work program, planned changes in the program can be analyzed prior to their implementation to determine the effects of the changes, thus improving the success rate of program changes.

A simplified CPN chart is shown in Figure 8.6.

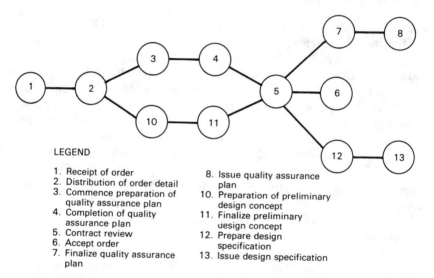

LEGEND

1. Receipt of order
2. Distribution of order detail
3. Commence preparation of quality assurance plan
4. Completion of quality assurance plan
5. Contract review
6. Accept order
7. Finalize quality assurance plan
8. Issue quality assurance plan
10. Preparation of preliminary design concept
11. Finalize preliminary design concept
12. Prepare design specification
13. Issue design specification

Figure 8.6 Critical path network.

From a quality auditor's point of view, the major advantage of this technique lies in the ability to use an existing set of documents as the analytical tool. The fact that the documents are already in use will assist the auditor in discussing the audit with other personnel during the planning phase of the CPN and the audit. The technique is extremely useful in conducting product quality audits, particularly for the first run of a major product. Moreover, its use of production schedules helps in the scheduling of audits. Where a procurement quality standard requires a schedule for use by customer auditors, a CPN chart will suffice. Thus the technique can serve both internal and external auditors.

The major disadvantage of the traditional CPN printout is the lack of visibility it offers to the controls, procedures, standards, etc., relating to the various activities it documents. However this shortcoming can be overcome by a minor modification to the computer program enabling it to provide a second listing of the activity legends giving the needed data. The two legend tables together then provide the quality auditor with data on both scheduling and quality control.

Although the above comments have discussed CPN in terms of computer printouts and CPN programs are readily available for most business-sized personal computers, CPN charts can be done manually. In fact, it is recommended that simplified CPN charts *be* done manually before automating the entire sequence of activities. In this way the skills and discipline necessary for producing CPN charts and understanding them can be developed.

8.5   Matrix Responsibility Chart

The matrix responsibility chart is a method of summarizing the actions to be taken on the various procedures that are used within an organization. It can also serve as a very effective index for a procedures manual. Where such a chart is available, it provides an excellent planning tool for conducting audits to determine compliance with procedures, i.e., conformity audit.

The technique uses a format along the lines of the one shown in Figure 8.7. Vertical columns are used for the procedures, etc., with a single column per policy, procedure, instruction, etc. Each horizontal line represents a function within the organization, e.g., marketing, sales, design, procurement, materials control, production, quality, etc. If different levels within a function have different responsibilities, it may be necessary to have more than one column for that function, possibly incorporating job titles.

The input of a function to a given procedure is shown by a symbol at the intersection of the line and column. The following are typical symbols used to denote various actions:

Symbol	Action
S	Responsible for starting or initiating an action, e.g., starting a form, movement, etc.
A	Mandatory approval of a form or activity. If more than one approval is required numbers after the letter A show their sequence.
I	Provided with an information copy of the form.
U	User of the procedure, instruction, etc.
W	Work to be done as the result of a procedure, instruction, etc.
Blank	No input or required action.

```
┌─────────────────────────────────────────────────────────────────────┐
│ Decommissioning _____ │
│   Service _____ │
│   Commissioning _____ │
│   Installation _____ │
│   Publications _____ │
│   Shipping _____ │
│   Packing _____ │
│   Inspection and test _____ │
│   Assembly _____ │
│   Feeder manufacturing _____ │
│   Manager production _____ │
│   Industrial engineering _____ │
│   Manufacturing engineering _____ │
│   Production control _____ │
│   Manager manufacturing planning _____ │
│   Stores _____ │
│   Materials verification _____ │
│   Receiving _____ │
│   Purchasing _____ │
│   Manager materials control _____ │
│   Design verification _____ │
│   Modeling _____ │
│   Design documentation _____ │
│   Software design _____ │
│   Project design _____ │
│   Manager design _____ │
│   Contract administration _____ │
│   Marketing and sales _____ │
│   Manager marketing _____ │
│   Quality engineering _____ │
│   Quality audit _____ │
│   Manager quality assurance _____ │
│   Accounting _____ │
│   Manager accounting _____ │
│   Executive _____ │
│   Procedure number _____ │
│   Procedure title _____ │
└─────────────────────────────────────────────────────────────────────┘
```

Figure 8.7   Matrix responsibility chart.

The development and use of this type of form can be helpful in ensuring that each and every input to a procedure, instruction, etc., is addressed during the planning stage of an audit. It is particularly helpful in deciding whether to use a function- or location-oriented quality audit. In cases where the organization being audited does not employ this technique, it is frequently useful for the auditor to have blank forms that he or she can fill in during the procedural check to summarize the activities to be covered.

The technique's major disadvantage is the lack of visibility it provides as to where the various documents involved fit into the particular product flow.

8.6   Function Tables

Function tables are a tabular version of function trees where once again an organization is broken down into its major units. Each of these units then has subsequent tiers of functions for which it is responsible.

By using a format similar to that shown in Figure 8.8, a system can be analyzed to determine its functional activities and their control and verification requirements. In reviewing an area, the first productive function or activity is noted in the Function column. The applicable specifications, drawings, standards, etc., are then listed in the second column. The applicable assurance activities are entered in the Assurance column, followed by any relevant procedures in the Reference Documents column. Normally, the productive function and the assurance function will be related on a one-to-one basis. However if more than one of either is involved, additional lines should be used in the correct sequence of occurrence.

This technique complements some of those described earlier in that it provides clear visibility for the relevant documentation relating to the productive and assurance functions. As an analytical tool, it has no particular advantages or disadvantages other than its ability to be entered in its complete form on a word processor for inclusion in other documents.

Figure 8.8 illustrates the use of this technique for the early phases of a quality audit project.

8.7   Ishikawa (Cause and Effect) Diagram

The Ishikawa diagram—also known as the cause and effect or fishbone diagram—was developed in Japan by Professor Ishikawa as an analytical tool. It is frequently used in quality circle activities.

The diagram was originally developed as a means of identifying all the interacting activities that contribute to the output of a given process by subdividing the process into the various components, subprocesses, environment, etc., that must be individually controlled to ensure the success of the overall process. It has become a valuable problem-solving tool since it allows an analyst studying a problem to consider all the possible contributory causes involved with the problem being reviewed.

The problem or process is entered in the head of the fishbone. In the traditional diagram, the major arms are identified as:

1. Machine influences
2. Material influences
3. Worker influences
4. Method influences

Function	Control documents	Assurance activities	Reference documents
1. Definition of customer needs	Product specification		
2.	Product quality plan	Definition of assurance activities	
3.	Product specification Product quality plan	Contract review	
4. Circulation of contract information	Work order		
5.		Quality audit	
6. Prepare design plan			
7. Preliminary design	Product specification		
8.	Preliminary design Product specification Product quality plan	Initial design review	
9. Detailed design	Product specification Product quality plan Product design plan Preliminary design Qualified parts lists etc.		
10.		Quality audit	
11.	Product specification Product quality plan Regulatory requirements Reliability analyses Maintainability analyses Availability analyses etc.	Verification tests	
12.	Product specification Product quality plan Product design documents Verification data	Design review	
13. Release for procurement and production			

Figure 8.8   Function tables.

Experience in various disciplines has shown that it can be advantageous to add one or more of the following arms:

1. Environmental influences
2. Regulatory influences
3. Transportation influences
4. Storage influences

In practice, each of the major arms are usually subdivided one or more times to show the various levels of activities. The subdivision can be carried through as many steps as necessary to reach the basic elements involved.

From a quality audit point of view, the diagram can be used effectively to identify all of the contributory elements of the system, process, product, etc., being evaluated. It is particularly useful where the audit is being conducted to identify the cause or potential cause of an already identified problem.

A typical Ishikawa diagram is shown in Figure 8.9.

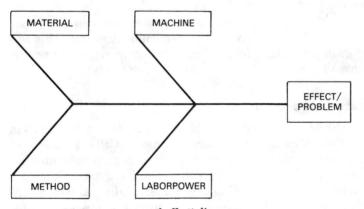

Figure 8.9 Ishikawa (cause and effect) diagram.

8.8 Recommendations

8.8.1 General

As indicated in the above paragraphs, each of the planning aids discussed has various strengths and weaknesses. This section will provide recommendations on the techniques that the author has found to work most successfully in various situations. For other people, the preferred techniques may be some of the others described in this chapter or entirely different techniques developed elsewhere.

8.8.2 Quality program improvement

In planning a quality audit to evaluate an existing quality program with a view to improving it, the ideal evaluation tool is the function tree. This technique permits the planner to develop an ideal system based on all of the applicable functions, as described in Section 8.2. In addition, it allows the addition of any special regulatory or standards requirements for the product line being audited. The result will be a function tree similar to that shown in Figure 8.3. Auditors or consultants specializing in this kind of activity can have such a diagram available in their data libraries, ready for customizing if necessary.

8.8.3 Auditing to a quality standard

When an evaluation is to be made against a quality program (system) standard, there is not much difference between the function tree and the decision flow chart techniques. Either can be readily adapted to cover all the elements required by the particular standard. Auditors involved in this kind of activity should use a standard working paper (checklist, etc.) for all audits involving a particular standard. This will help to maintain a constant level of expectation in regard to performance against that standard.

8.8.4 Product or service audit

A product or service audit will be best served by either the decision flow chart or the critical path network technique. CPN is somewhat stronger and more robust. However, this advantage barely justifies the additional effort and facilities needed to develop a network for use by only the auditor. If CPN is being used for scheduling as well as audit purposes, every effort should be made to use the same network for both. The joint use will strengthen each application through the additional inputs and verifications generated.

8.8.5 Investigatory audits

For this type of audit, designed to investigate a given process, product, etc., for actual or potential problems, the Ishikawa diagram comes into its own. It has been found to be effective for both evaluation and reporting purposes.

9

Working Papers

9.1 Introduction

The definition of "working papers" is virtually the same in the American and Canadian quality audit standards ANSI/ASQC Q1-1986 and CAN-CSA-Q395-1981, respectively. Working papers are defined in the ASQC standard as:

> **Working Papers**: These are all of the documents required for an effective and orderly execution of the audit plan. By format and content they describe the scope and approach of the audit assignment and its operational elements.

Thus the term includes such documents as:

1. Quality audit instructions or procedures (discussed in Section 9.2)

2. Planning documents (discussed in Section 9.3)

3. Working papers or checklists (discussed in Section 9.4).

4. Auditor reporting documentation (discussed in Chapter 14)

Although all these are *working* documents, I believe they should be prepared in a professional manner that will portray the desired quality image of the auditing organization, i.e., they should never be just penciled notes on loose sheets of paper.

The majority of the techniques and working paper layouts discussed in this chapter are suitable for both external and internal quality audits.

9.2 Quality Audit Instructions or Procedures

These documents are used to issue operating instructions to the auditors. They include procedures for each of the areas discussed in this text, normally with one procedure per technique, form, etc. Typical techniques described by them would be:

1. Use and selection of planning documents
2. Preparation of audit working papers
3. Use of sampling plans
4. Implementation of the various types of audits
5. Recording of observations
6. Reaction to anomalies or nonconforming observations

The instructions may be prepared by the auditors or their supervisor. Regardless of who originates them, the documents should be subject to approval by the manager or head of the auditing organization.

Typical audit instructions applicable to location-oriented internal quality audits are shown in Appendixes 9C and 9D. Appendix 9C relates to a manufacturing activity, while Appendix 9D relates to an administrative one. These instructions tell the auditor how to implement various of the working papers involved in each instance. This type of working paper is normally laid out in the manner shown in Figure 9.4c through f. In that form, the sample size and the results of the sample are clearly visible.

As with all procedures or instructions within an organization, care should be taken to ensure that audit instructions are prepared in a way and use a level of language understood by their intended readers. Doing so will not present as great a challenge for audit instructions as for more general operating instructions because of the narrower spread in education and language fluency that exists among auditors than among workers in general. Nevertheless, wherever possible, flow diagrams or other diagrammatic presentations are preferable to words alone.

It is frequently advisable to issue audit instructions interpreting each quality standard used as a reference document—those covering systems, products, processes, services, etc. This helps ensure a common viewpoint among all the auditors, and is of great assistance to new auditors joining the organization. Such interpretive instructions will undoubtedly be expanded upon as the number of audits the organization undertakes increases. Thus each one will become a document collating the various interpretations of each standard, and provide an up-to-date history of the kinds of problems encountered in audits related to it. It can be useful to note the dates and locations of the auditing activities that contributed to each particular interpretation.

In designing the layout of an auditing instruction document, it is recommended that the starting point be each significant paragraph in the standard. This should include all numbered paragraphs, and the numbering system should be extended, if necessary, to cover any significant paragraphs not numbered in the standard itself. As new interpretations arise, they should be entered under the appropriate paragraph number and heading in chronological sequence of occurrence, from the latest (directly under the heading) to earliest. Each of these interpretations should be given its own serial number, starting with 01 for the earliest. In other words:

1. Begin with the paragraph number and title from the standard.

2. Directly below this enter the serial number and a description of the most recent interpretation, along with the date(s) and location(s) of the auditing activities that contributed to it.

3. Enter the serial number and a description of the second most recent interpretation, along with date(s) and location(s) of the relevant audit activities.

4. Repeat step 3 for each interpretation, going chronologically backward toward the earliest interpretation.

5. Enter the relevant data for the earliest interpretation. The serial numbering system for interpretations starts here, with 01.

Quality audit instructions should project a quality image compatible with that desired for the organization. A simple way of achieving this is to have all the instructions prepared on a word processor, with their layout designed to match that of followup pages used for external correspondence. The latter usually identify the organization with a logo and/or the company name. Each auditor should have a set of quality audit instructions, bound in a suitable cover. Distribution outside the auditing organization should be restricted as far as possible; in regard to auditees, preferably only the chief executive officer should receive a copy, and then only if requested.

9.3 Planning Documents

Each of the planning documents (discussed in Chapter 8) selected for use should be covered by an audit instruction describing the document's advantages and disadvantages and how to use it. If a variety of methods are available, an overall instruction on how to select the appropriate document should be included. This instruction would cover the advantages and disadvantages of each planning document as part of the selection process.

Program learning logic can greatly simplify this type of selection guide. That is:

1. Is the audit being conducted to investigate a problem?

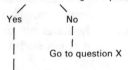

```
              /        \
           Yes          No
            |            |
            |            |
            |         Go to question X
            |
```

2. Is it evaluating a system, product, process, or service for
   compliance with a given standard?

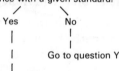

```
              /        \
           Yes          No
            |            |
            |            |
            |         Go to question Y
            |
```

3. Is it evaluating the overall quality system?

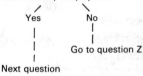

```
              /        \
           Yes          No
            |            |
            |            |
            |         Go to question Z
            |
       Next question
```

The first question separates quality audits involved with problem solving from those evaluating quality systems in their various forms. Question X would then be the first question having to do with problem solving. The second question separates audits involved with program improvement from conformity quality audits. Question Y would be the first question in the improvement sequence. Similarly the third question separates conformity audits into those covering a complete system and those covering specific products, processes, etc.

9.4 Working Papers

9.4.1 Introduction

In this section, the term "working papers" is used to cover the variety of formats that can be used as memory prompters for the auditor and to record the results of the examination of each element being audited. It has a long history of use in the accounting profession to cover those documents prepared to ensure that auditors cover every applicable point and provide space for recording the results. The term "checklist"

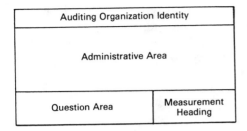

FIG. 9.1 Typical layout for quality audit working papers.

is sometimes applied to these documents. Checklists traditionally have only two columns for answers to auditing questions, usually labeled "Yes" or "No." However, I feel "yes" or "no" alone is too limited an answer, and prefer a different approach.

These documents have four basic areas, with any technique applicable in one area compatible with any technique applicable in the others. Most areas apply to virtually any type of audit. They are (see Figure 9.1):

1. Auditing organization identity
2. Administrative data
3. Questions
4. Measurement of the results

9.4.2 Auditing organization identity

In laying out the working paper format, the section marked "Auditing Organization Identity" should clearly indicate the name of the auditing organization. A simple way of achieving this is to photocopy the data onto an external correspondence followup page. Such pages normally carry the organization name and/or logo. If auditing is a subdepartment of the organization, the subdepartment title should also be included:

Productivity and Quality Improvements, Inc.
Quality Audit Department

9.4.3 Administrative area

As shown in Figure 9.1, the administrative area normally covers the full width of the paper below the organization's name, the questions area takes up most of the space under that, and the measurement area is located to the right of the questions area, also under the administrative area.

As shown in Figure 9.2, the administrative area provides headings for all of the data necessary to identify the audit:

1. What type of audit is being conducted
2. Who is being evaluated
3. Why the audit is being conducted
4. Who the auditor is
5. What the reference standards are
6. When the audit is being conducted

The following notes apply to the headings shown in Figure 9.2:

1. *Sheet No.*   This provides the page numbering for the questions applying to a major function.

2. *Major Function.*   This identifies the major function involved with the particular type of audit, e.g., for a quality system audit, the major function would be selected from the following: executive action, marketing, design, materials control, production, sales, installation, commissioning, decommissioning, etc.

3. *Area.*   This identifies the particular subfunction of the major function being audited by subdepartment number or name. When several tiers or levels of subfunctions are involved, this section should identify the lowest tier.

4. *Type of Audit.*   This describes the type of audit being undertaken, i.e., systems, product, process, service, suitability, conformity, etc.

5. *Reference Standard.*   This identifies the system standard forming the reference document for the audit. If process, product, or other lower-level standards are applicable to the area covered by the working paper, they should also be noted, subject to space limitations.

```
┌─────────────────────────────────────────────────────────┐
│              Auditing Organization Identification          │
├─────────────────────────────────────────────────────────┤
│              Quality Audit Working Papers                  │
│                   Sheet No. _____                     │
│                                                            │
│  Major Function _____   Audit No. _____  │
│  Area _____       Date _____  │
│  Type of Audit _____    Auditor _____  │
│  Reference Standard _____  │
│                                                            │
└─────────────────────────────────────────────────────────┘
```

FIG. 9.2   Typical layout of administrative area on working papers.

6. *Audit No.*   The auditing techniques described in this book permit the storing of the data derived in a data processing system, with the "Audit No." serving as the recall or sorting character. But the recommended use of this line is somewhat different for external and internal quality audits.

For external quality audits, it is recommended that each supplier be coded for identification, and then the audits being carried out on each supplier be numbered sequentially. In this way, the supplier's name does not appear on the working papers, and the confidentiality of the activity is safeguarded.

For internal quality audits, it is recommended that the audits simply be numbered sequentially. If a large number are being conducted, it may be advisable to use the month as a prefix; sequential numbers should then start with 1 for each month.

7. *Date.*   This is self-explanatory. However, it is recommended that the date be shown in the numerical sequence of year, month, day of month. This permits easier data processing.

8. *Auditor.*   This should include the assigned auditor's name (typed) and the signature of the actual auditor on completion of the audit.

9.4.4   Question area header

The question area contains the header or title block and the questions to be used during the audit. Typical headers are shown in Figure 9.3.

In most audits, the reference against which an operation or activity is compared will be a paragraph of either the reference standard or the applicable reference procedure. The title of this reference paragraph will be shown in the page header.

Since the first queries tend to be exploratory, the auditor must be able to develop suitable followup questions. However, care must be

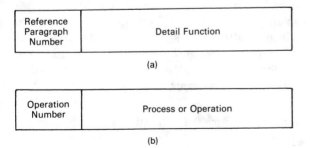

FIG. 9.3   Typical question area headers on working papers: (*a*) for a general quality audit; (*b*) for a process quality audit.

taken in phrasing the exploratory questions, as they will set the climate for the remainder of the investigation.

All questions should be phrased in an unbiased way and so as to permit constructive and useful answers. Every attempt should be made to avoid questions which will put the other party on the defensive.

Loaded questions have no place in a quality audit. Fundamentally, a loaded question is one where the auditee is placed in a losing position regardless of the answer. An example of a loaded question, taken from the vaudeville stage, is: "Have you stopped beating your wife?" If the answer is "yes," the responder is obviously a brute, albeit a reformed one. If the answer is "no," the responder is an unreformed brute. There is no answer for the civilized individual who does not beat, and has never beaten, his wife. This example may seem oversimplified. However, during many years of involvement in audits, as an audit observer, a client, and an auditee, I have found many occasions when questions about quality systems have been phrased in similar no-win terms.

Another type of question to be avoided is one that solicits an answer in the form of "No, but..." or "Yes, but...," because it tends to lead to a defensive situation.

In developing investigative techniques, a quality auditor must seek ways to maintain proper and constructive interpersonal relationships with those being audited. Any tendency toward an inquisition or vendetta will destroy the relationships so essential for a successful quality audit. An auditor is a peaceful investigator, not a modern Torquemada.

Examples of typical questions are given below in Section 9.4.6, "Typical Quality Audit Working Paper Questions."

9.4.5 Measurement area

9.4.5.1 Introduction. There are a number of variants which can be found in the measurement area of working papers. Typical measurement area headers are illustrated in Figure 9.4a through f. Two of these, shown in parts a and b, are judgmental and qualitative in nature; the headers shown in parts c through f, on the other hand, are quantitative in nature. Since quality audits should be quantitative as far as possible, the headers in parts c through f are preferred. In all cases, the sample size to be used for verifying statements should be entered at the planning stage. Selection of this is discussed in Chapter 10.

9.4.5.2 OK/NOK measurement. Figure 9.4a shows a typical header for a checklist, i.e., for a series of questions to determine the presence or absence of a particular process or control feature.

This type of list was common in the early days of quality auditing. As auditing technique developed, though, shortcomings were rec-

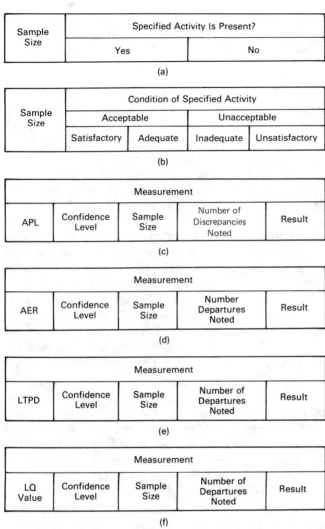

Sample Size	Specified Activity Is Present?	
	Yes	No

(a)

Sample Size	Condition of Specified Activity			
	Acceptable		Unacceptable	
	Satisfactory	Adequate	Inadequate	Unsatisfactory

(b)

	Measurement			
APL	Confidence Level	Sample Size	Number of Discrepancies Noted	Result

(c)

	Measurement			
AER	Confidence Level	Sample Size	Number Departures Noted	Result

(d)

	Measurement			
LTPD	Confidence Level	Sample Size	Number of Departures Noted	Result

(e)

	Measurement			
LQ Value	Confidence Level	Sample Size	Number of Departures Noted	Result

(f)

FIG. 9.4 Typical headers for the measurement area on working papers: (a) for a checklist; (b) for graded performance; (c) for an acceptable performance level; (d) for an acceptable error rate; (e) for lot tolerance percent defective; (f) for limiting quality level (LQL) sampling plan.

ognized, and there is now a tendency to avoid using this form of audit instruction.

There is, however, a place for it in the menu of audit measurement techniques. The yes/no type of questionnaire is very effective in carrying out a conformity quality audit to evaluate the conformance of a critical process to detailed operating instructions. It is also effective for determining the compliance of quality system documentation with quality standards requiring specific functions. It should not be used for a conformity audit dealing with more general processes or procedures, though, as simple yes/no situations seldom arise in practice.

9.4.5.3 Graded measurement. Figure 9.4b allows for the kind of intermediate assessment where a minor departure from the reference standard is noted. In such cases, the terms "Satisfactory" and "Unsatisfactory" are self-explanatory and cover those situations where the evaluation has determined conditions to be clearly acceptable, or clearly unacceptable with prompt corrective action required.

The terms "Adequate" and "Inadequate" are used to cover those minor divergences from requirements where corrective action is required but on a lower priority than for an "unsatisfactory" condition. Thus they are less clear, and need some explanation. "Adequate" is used where a questionable condition is noted and there is a tendency toward an unacceptable condition, but the dividing line has not been crossed. Corrective action in this case is intended to prevent problems from arising. "Inadequate" is used where the departure from requirements is minor and unacceptable, but not sufficiently serious to cause the condition to be classed as "unsatisfactory." Corrective action, in this case, is a higher priority than for an "adequate" situation, but not as high as for an "unsatisfactory" one. The dividing lines between the gradations will depend on the discipline and processes concerned.

9.4.5.4 Statistical measurement. Figure 9.4c, d, e, and f illustrate headers that are quantitative in nature and based on statistically sound sampling plans (discussed in Chapter 10). Suffice it at this point to say that in each case the audit planner must evaluate each requirement for criticality and assign some acceptable level of performance. The planner must also assign a confidence level at which the sampling plan can be depended on to detect any substandard performance. From these two values, the sample size can be determined and entered on the form. The selected sampling technique will determine whether the header shown in Figure 9.4c, d, e, or f is used. Guidance on the selection of sampling factors is given in Chapter 10.

During an audit, the auditor enters on the form the number of departures from the requirements observed. Once the audit is completed, each entry showing discrepancies is reviewed. Each line should be evaluated as discussed in Chapter 12, and the result or acceptability determined. Since corrective actions must be prioritized to a degree, the classification of the results as satisfactory, adequate, inadequate, or unsatisfactory may suffice. As will be noted in Chapter 10, the sampling plans recommended are fundamentally zero-acceptance number attribute sampling plans; thus any departure from requirements noted implies the need for corrective action at some level.

9.4.6   Typical quality audit working paper questions

9.4.6.1   General. In considering typical questions for inclusion on working papers, it must be remembered that most sets of questions can be used with virtually any combination of the various headers shown in Figures 9.1 through 9.4. It should also be pointed out that values for performance levels or confidence levels have not been included in those illustrations since they will vary according to the discipline involved, the application, etc.

9.4.6.2   Quality system improvement questionnaire. Appendix 9A gives a set of basic questions for use in determining the status of and providing the basis for improvements to the quality system of an organization involved with designing and manufacturing products for the high tech marketplace. It has been prepared to match the function tree shown in Figure 8.3 rather than with a particular procurement quality standard in mind. However, it could be used with most of the major procurement standards reviewed in Chapter 1.

9.4.6.3   Questionnaire for an audit based on ISO 9001-1987. Appendix 9B provides a set of working papers for use with an audit to determine the suitability of a quality system with respect to ISO 9001-1987. In this case, a set of questions has been developed for each of the 17 activities covered by the standard. However, an examination of these sections shows that several apply to more than one function within an organization.

If we consider that each function within an organization involves both a productive act and a verification act, i.e., that titles are functional and not necessarily departmental, a form such as that shown in Figure 9.5 can be developed. Thus the package of audit working papers for one of the organizational elements shown at the top of Figure 9.5 would include the forms for each ISO element (left-hand column) checked off below. For example, the package for auditing the design activity would consist of the working papers shown in 9B.1–9B.12 (with the possible exception of 9B.8), plus 9B.14 and 9B.16. Any of the functions—marketing, design, materials control, production, service, installation, and quality—can be subdivided into more than one department or functional entity. In those cases, duplicate sets may be required for each subunit.

The basic questions given in Appendix 9B can be used for both suitability and conformity quality audits; however the answers would be checked against the reference standard in the first case and against the organization's procedures as well as the standard in the second.

Locations subject to quality audits are not shown on Figure 9.5, but packages of working papers in their cases are similar. For example,

ISO function	Organizational element						
	Mktg.	Des.	Mat.	Prod.	Serv.	Inst.	Qual
Management (Appendix B1)	✔	✔	✔	✔	✔	✔	✔
Quality system (Appendix B2)	✔	✔	✔	✔	✔	✔	✔
Contract review (Appendix B3)	✔	✔	✔	✔	✔	✔	✔
Design (Appendix B4)	—	✔	✔	—	—	—	✔
Document control (Appendix B5)	✔	✔	✔	✔	✔	✔	✔
Purchasing (Appendix B6)	—	✔	✔	✔	—	✔	✔
Identification and traceability (Appendix B7)	—	✔	✔	✔	—	✔	—
Process control (Appendix B8)	—	✔	✔	✔	✔	✔	—
Inspection and test (Appendix B9)	—	✔	✔	✔	✔	✔	—
Metrology (Appendix B10)	—	✔	✔	✔	✔	✔	—
Nonconforming material (Appendix B11)	✔	✔	✔	✔	✔	✔	✔
Handling, storage, packaging, and delivery (Appendix B12)	✔	✔	✔	✔	✔	✔	—
Quality audits (Appendix B13)	—	—	—	—	—	—	✔
Training (Appendix B14)	✔	✔	✔	✔	✔	✔	✔
Servicing (Appendix B15)	✔	—	✔	—	✔	—	—
Statistical techniques (Appendix B16)	—	✔	✔	✔	✔	✔	✔
Installation (Appendix B17)	✔	—	✔	—	—	✔	—

FIG. 9.5 Application of ISO 9001-1987 working papers to an industrial organization.

working papers for an audit of a manufacturing location would consist of the following forms:

1. Management (Appendix 9B.1)
2. Quality system (Appendix 9B.2)
3. Documentation control (Appendix 9B.5)
4. Identification and traceability (Appendix 9B.7)
5. Process control (Appendix 9B.8)
6. Inspection and test (Appendix 9B.9)
7. Metrology (Appendix 9B.10)
8. Nonconforming material (Appendix 9B.11)
9. Handling, storage, packaging, and delivery (Appendix 9B.12)
10. Training (Appendix 9B.14)
11. Statistical techniques (Appendix 9B.16)

Obviously not every question shown in the appendixes will apply in every audit situation or location. However, some of the questions in each will apply in every instance. Thus the audit instructions must provide some guidance on the editing of the overall questionnaire. It should also be noted that certain activities not listed in ISO 9001-1987 could become the audit subjects and sheets would be required for these. Typical of such areas are: worker awareness of an organization's quality system and individual participation in it, employee involvement with quality circles or other employee participation programs, the effects of automation in certain areas, housekeeping, etc.

9.4.6.4 Manufacturing area—detailed instructions. Appendix 9C provides a set of questions for a location-oriented quality audit as applied to a single recognized manufacturing group, e.g., cost center, group, foreman, etc. As part of a repetitive internal audit, the sample sizes on a single visit may be smaller than if the activity is being audited as part of a system evaluation.

9.4.6.5 Administrative area—detailed instructions. Appendix 9D provides a set of detailed questions that can be applied to the audit of a typical administrative area, in this case the preparation and typing of a purchase order.

9.4.6.6 Service industry. The preparation of working papers for an audit relating to a service industry has some special connotations, but in general follows the same ground rules already cited.

External quality audits to determine the suitability of a service industry against some national or international standard tend to be specialized as they usually involve conformance to some government regulation, e.g., audits of hospitals, convalescent homes, senior citizen homes, etc. In these cases, the working papers are prepared following the appropriate regulations, statutes, etc., to cover each element required. The working papers are checked against the policies, manuals, procedures, instructions, etc., that document the organization's quality system. If the documentation is satisfactory, further audit sheets are prepared, similar to those shown in Appendixes 9C and 9D, to be used in checking the actual operations against the procedures. Decision sampling is a key technique in these fields. As with audits of manufacturing organizations, once the working papers have been prepared, the same ones should suffice for any future audits based on that same set of regulations.

For unregulated services, audits fall into two categories:

1. Those used to determine the degree of conformance to and effectiveness of an organization's own quality program

2. Those used to determine the quality program's effectiveness from the customer's point of view

Working papers for the first category are derived from the organization's documentation as described above and used by auditors either working in the facility full time or out of a central office to cover a number of facilities. (Auditors usually work full time at facilities such as banks, hotels, etc.; auditors usually work out of central offices for such chain businesses as fast food outlets, repair shops, travel agencies, etc.) They are prepared the same way as any other working papers for determining conformance to quality system documentation. Audits in both categories use decision sampling as one of the means of evaluation.

The auditors carrying out an evaluation of a quality program's effectiveness from the customer's point of view will not be known to the facilities and individuals being audited. They may be employees of the organization being audited or under contract to a third party. The working papers for such audits will be prepared from the point of view of customers and the service that customers expect to receive. A typical set of questions relating to service for an airline traveler is shown in Appendix 9E. Working papers of this sort cover the normal sequence of events for the activities being audited from initial contact through completion of the transaction.

Appendix 9A

Typical Working Papers
for Evaluating
a Quality System
for Quality and Productivity
Improvement

APPENDIX 9A.1   Typical Set of Questions for Use in Quality System Evaluation—Executive Action

Reference paragraph number	Detailed function
E-1	Do you have a corporate quality policy? If yes: How is it defined? How is it promulgated? To whom is it promulgated? Is it refined to provide effective quality policy statements for each major operating function?
E-2	Please describe your organization and in particular the interrelationships, responsibilities, and accountabilities of the various levels of management. Supplementary questions if not answered as part of the above question: May we have a copy of your organization chart? How are the responsibilities and accountabilities defined, in general and for individuals, and promulgated? Are all personnel aware of the quality policy and their individual contribution to it?
E-3	How do you demonstrate, on a continuing basis, your commitment to the quality policy, with any supporting documentation?
E-4	Is your quality program or system defined? If yes: How is it documented? *e.g., quality manual, procedures, instructions, quality, inspection or test plan, other_____* Who prepares the documents? Who reviews or approves the documents? How are changes introduced and controlled? How do you customize your general program to the specific needs of a contract, customer, or product? How are personnel made aware of these documents? If no: How are personnel aware of their input to achieving the corporate quality policy?
E-5	How is conformance to your policies, procedures, etc., determined? If departures are noted, who is responsible for assigning the responsibility for any required corrective action? Who approves the method and verifies any corrective action required?

APPENDIX 9A.2   Typical Set of Questions for Use In Quality System Evaluation—
Marketing

Reference paragraph number	Detailed function
M-1	How are the customer performance, quality assurance, and other requirements established? Supplementary question: Are tenders and contracts treated in a similar manner?
M-2	How is this data circulated to all parties concerned?
M-3	How do you determine if the data circulated provides all the necessary information for the necessary tasks to be performed and verified? *e.g., product definition, tender review, contract review, other_____*
M-4	Is marketing involved in any of the design and production control cycles? *e.g., design reviews, quality reviews, configuration changes, disposition of nonconforming material, other_____*
M-5	How is advertising reviewed or related in any way to the design and production controls?
M-6	How are marketing contributions to the quality system/program documented?

APPENDIX 9A.3   Typical Set of Questions for Use in Quality System Evaluation—Design

Reference paragraph number	Detailed function
D-1	How do you ensure consistent and verified design practices are used by all designers? *e.g., standard practices for hardware and software design, standard parts lists complete with application notes, procedures, other*_____
D-2	How are reliability, maintainability, and availability (RAM) requirements handled from the design and verification points of view? *e.g., quality or reliability plan, product specification, other*_____
D-3	How is the design evaluated or verified against the marketplace requirements? *e.g., verification tests, computer model, other*_____
D-4	How is configuration defined and controlled?
D-5	What controls apply to metrology during the design and verification phases?
D-6	How is metrology defined and controlled for use during the production or manufacturing phase?
D-7	Are design reviews held? If so: When are they held and what form do they take? How are the requirements documented?
D-8	How do you confirm conformance to the procedures and other documentation?

APPENDIX 9A.4   Typical Set of Questions for Use in Quality System Evaluation—Materials Control

Reference paragraph number	Detailed function
MC-1	Please describe your procurement cycle. Supplementary questions: How are new suppliers selected? *e.g., vendor appraisals, qualified supplier lists, past performance, CASE, other_____* How are any quality terms and conditions determined? *e.g., product or contract inspection plan, procedures, contract data, other_____* How is the configuration of the purchased goods defined and controlled? *e.g., procedures, product or contract quality plan, other_____*
MC-2	Are all of the above covered by some form of documentation? *e.g., procedures, purchasing instructions, product or contract quality plan, other_____*
MC-3	How is expediting controlled to prevent degradation of the product? *e.g., procedures, other_____*

MC-4	How is a request from a supplier for permission to deviate from a purchasing requirement handled?
	Supplementary questions:
	  Who are involved in the disposition?
	  How is the criticality assessed?
	  How is disposition agreed on?
	  How is disposition reflected to the supplier?

MC-5	How are materials controlled on receipt to maintain identity and prevent degradation?
MC-6	How is the acceptability of purchased goods and services determined? *e.g., source surveillance, vendor-generated data, vendor quality system approval, incoming or receiving inspection, on application or usage, other_____*

**APPENDIX 9A.4   Typical Set of Questions for Use in Quality System Evaluation—
Materials Control** (*Continued*)

Reference paragraph number	Detailed function

MC-7	How is the flow of materials to and from stores controlled? Supplementary questions:
	    Do the controls provide the necessary batch or lot controls?
	    Can materials bypass the control system in any way?

MC-8	How is material integrity maintained in stores? *e.g., basic identity, environmental protection, shelf life, batch or lot identification, storage containers, segregation by project, other_____*
MC-9	What do you have in the way of special equipment or facilities to prevent damage of materials during handling?
MC-10	How are the material control rules disseminated to ensure employee awareness and compliance? *e.g., verbal instructions, procedures, written instructions, other_____*

Reference paragraph number	Detailed function
P-1	Please describe your production process along with its controls and verifications. Supplementary questions: Do the operators have the means of confirming that their completed operations conform to the requirements? What training do the operators receive in regard to methods and working standards? Where do operators find information about an acceptance standard if they are not sure of the requirement? Are all operations covered by written instructions?
P-2	Do you carry out process capability studies, and are these used in the selection of processes?
P-3	Are process control charts used? Supplementary questions: Where are they posted? Who plots the data? Who takes corrective action, if required?
P-4	How is metrology controlled and calibrated? *e.g., positive recall, all gauges calibrated, calibration status, other___*
P-5	How are production documents controlled to ensure correct configuration?
P-6	How is material identity and traceability maintained?
P-7	What handling facilities are provided to prevent damage to materials prior to, during, and after the production phase?
P-8	How are the production requirements verified? *e.g., in-process inspection, process control, last-point inspection or test, final inspection or test, product approval testing, periodic test and inspection, next-operator inspection, supervisory check, other___*
P-9	How do you review any patterns of nonconformity that may have developed? *e.g., quality reviews, production meetings, other___*
P-10	What happens when nonconforming materials are discovered? *e.g., segregation, verification, identification, documented, review, disposition, corrective action, other___* Supplementary questions: Who are involved in the review and disposition? What are the different dispositions and the actions after each? Who determines and approves the resulting corrective action?
P-11	Are production and verification activities documented?

APPENDIX 9A.6   Typical Set of Questions for Use in Quality System Evaluation—Packing and Shipping

Reference paragraph number	Detailed function
PS-1	How do you determine whether the packing method is controlled by the contract or the supplier? *e.g., contract information, contract quality assurance plan, contract production plan, other_____*
PS-2	If provided by you, how is the packing method determined and designed?
PS-3	How is the need for verification tests determined and who is responsible for performing them?
PS-4	How do you determine the necessary documents for inclusion?
PS-5	How is the method of shipping determined?
PS-6	How are the various packing and shipping requirements and controls defined? *e.g., verbal/written instructions, procedures, operator discretion, other_____*

APPENDIX 9A.7   Typical Set of Questions for Use in Quality System Evaluation—Quality Organization

Reference paragraph number	Detailed function
Q-1	Is the management representative designated to answer quality queries sufficiently knowledgable in the field?
Q-2	How does the quality function monitor performance of the various activities? *e.g., test and inspection, internal and external quality audits, decision sampling, data review, surveillance, outside services, other___*
Q-3	Who certifies the goods or services to the customer? Supplementary questions: What is the basis for certification? Does the individual certifying have the required accreditations? *e.g., professional and technical qualifications meeting the applicable regulations, contract requirements, etc.___*
Q-4	Does the quality organization provide the necessary statistical support? *e.g., sampling plans, statistical analyses, design of experiments, process or quality control charts, other___*
Q-5	Does the quality organization provide visibility on performance against quality standards? *e.g., customer complaints, quality cost reports, nonconforming material patterns, internal and external audit reports, other___*
Q-6	Does the quality organization provide visibility on the changes occurring in national and international quality standards?
Q-7	Are written guidelines provided on the various specialized techniques? *e.g., RAM procedures, statistical techniques, selection of quality standards, other___*

Appendix 9B

Typical Working Papers for Evaluating a Quality System based on ISO 9001–1987, *Quality Systems—Model for Quality Assurance in Design/Development, Production, Installation and Servicing*

APPENDIX 9B.1   Typical Set of Questions for Use in Quality System Evaluation based on ISO 9001-1987—Management

Reference paragraph number	Function
	How is management commitment to the quality system demonstrated?
4.1.1	Do you have defined and documented quality policies, corporate and departmental?
4.1.2.1	Defined and documented responsibility and authority of and the interrelationship between those who: 1. Manage work affecting quality 2. Perform work affecting quality 3. Verify work affecting quality
4.1.2.1	Defined and documented organizational freedom and authority to: 1. Initiate action to prevent the occurrence of product nonconformity 2. Identify and record any product quality problems 3. Initiate, recommend, or provide solutions through designated channels 4. Verify the implementation of solutions 5. Control further processing, delivery, or installation of nonconforming product until the deficiency or unsatisfactory condition has been corrected
4.1.2.2	Documented identification of verification requirements and provision of adequate resources including trained personnel for all verification activities: 1. Documented independence of those performing verification activities from those having direct responsibility for the work being performed **Note:** Verification activities include, but are not limited to, inspection, test and monitoring of the design, production, installation, and servicing processes, and/or product design reviews and audits of the quality system, processes, and/or product.
4.1.2.3	Documented appointment of a management representative who, irrespective of other activities, shall have defined authority and responsibility for meeting ISO 9001-1987 requirements
4.1.3	Documented procedures and records for periodic management review of the suitability and effectiveness of the quality system pertinent to the organization's activities and ISO 9001-1987: 1. Review conducted by members of the management team 2. Review conducted by independent external quality auditors 3. Review conducted by independent internal quality auditors

APPENDIX 9B.2   Typical Set of Questions for Use in Quality System Evaluation based on ISO 9001-1987—Quality System

Reference paragraph number	Function
4.2	How is the quality system documented?
	1. Quality manual defining responsibilities, authority, and accountability of functions and/or personnel affecting quality
	2. Procedures and instructions to define the methods of performing all aspects of the work affecting quality by those:
	a. Managing work
	b. Performing work
	c. Verifying work
4.2	Who is responsible for the following activities?
	1. Preparation of a quality manual
	2. Preparation of procedures and instructions
	3. Preparation of contract or product quality plans
	4. Identification and acquisition of:
	a. Control equipment
	b. Process equipment
	c. Verification equipment
	5. Identification and training, if necessary, of personnel for:
	a. Performing processes
	b. Performing verification tasks
	6. Updating and improving techniques for:
	a. Production
	b. Verification
	7. Determining measurements and processes outside the in-house capabilities of the organization
	8. Determination and clarification of all acceptance criteria
	a. Determining the compatability of the design, production process, installation, and inspection and test procedures, and applicable documentation
4.16	9. The identification, preparation, and retention of quality records
4.2	How is implementation of the quality system organized, documented, and verified?
	1. Organization charts
	2. Job descriptions
	3. Supervision
	4. Quality audits

APPENDIX 9B.3 Typical Set of Questions for Use in Quality System Evaluation based on ISO 9001-1987—Contract Review

Reference paragraph number	Function
4.3	How does the supplier define, establish, and maintain procedures for determining customer's needs and disseminate these needs throughout the organization? 1. Product definition 2. Dissemination 3. Contract review
4.3	Do the contract review procedures define or provide for: 1. A clear indication of who shall participate and the expected inputs and outputs? 2. Ensuring an adequate definition and documentation of the requirements? **Note:** The definition should be in quantitative measurable terms. 3. A clear understanding that the supplier can demonstrate the capability of meeting the contractual requirements? 4. Ensuring that any differences from the requirements of the tender are resolved?

Reference paragraph number	Function
4.4.1	Has the supplier established, documented, and maintained procedures and instructions to control and verify the design of the product concerned to ensure specified requirements are met?
4.4.2	Does the design plan procedure require the identification of who is responsible for each design and development activity?
4.4.2	Do the procedures require updating the design plan as the design evolves?
4.4.2	Do the project plans demonstrate compliance with the design plan objectives and the requirements of the procedures?
4.4.2.1	Do the procedures adequately require the assignment of qualified personnel, supported by the necessary resources?
4.4.2.2	Are the organizational interfaces between different groups defined satisfactorily?
4.4.2.2	How is intergroup information documented, transmitted, and regularly reviewed?
4.4.3	How are the design inputs for the product identified and documented and their selection reviewed for adequacy?
4.4.3	How are ambiguous or conflicting requirements resolved with those responsible for originating these requirements?
4.4.4	Are the design outputs documented and expressed in quantitative terms of requirements, calculations, analyses, etc.?
4.4.4	Is there evidence that design outputs: 1. Meet the design requirements? 2. Contain or reference acceptance criteria? 3. Conform to the appropriate regulatory requirements? 4. Identify those characteristics of the design that are crucial to the safe and proper operation of the product?
4.4.5	Do the procedures allow for the planning, establishment, documentation, and assignment of qualified personnel for the verification of the design?
4.4.5	Do the procedures for design verification establish the means of confirming that the design output meets the requirements of the design input by such control measures as: 1. Holding and recording design reviews? 2. Undertaking qualification tests and demonstrations? 3. Carrying out alternative calculations or mathematical models? 4. Comparison with an established, proven design? 5. Other means?_____
4.4.6	Are procedures established and maintained for the identification, documentation, and appropriate reviews and approvals of all changes and modifications?

APPENDIX 9B.5   Typical Set of Questions for Use in Quality System Evaluation based on ISO 9001-1987—Document Control

Reference paragraph number	Function
4.5.1	What procedures are established, documented, and maintained to control the preparation, approval, issuing, and modifying of all documents and data required by the contract and ISO 9001-1987? *e.g., marketing documentation, i.e., contract, product definition, tender and contract review records, dissemination documents, etc.; design documentation, i.e., specifications, drawings, computer programs, verification records, design reviews, etc.; materials control documentation, i.e., purchase orders, verification records, etc.; production documentation, i.e., production methods, production records, verification records, etc.; packing and shipping documentation, i.e., packing design and verification, shipping documents, quality certificates, etc.; service documentation; quality documentation, i.e., audit records, failure analyses, corrective action requests and followup, etc.; other_____*
4.5.1	Do these procedures ensure: 1. That pertinent documents are available to personnel at all locations? 2. That all obsolete documentation is promptly removed from all operations?
4.5.2	Do the procedures ensure that all changes to documents are at least reviewed and approved by those functions approving the initial document?
4.5.2	Is there a master list(s) providing visibility on the current revision status of all controlled documents?
4.5.2	Do the procedures require the reissuing of documents after a practical number of changes have been made?

APPENDIX 9B.6   Typical Set of Questions for Use in Quality System Evaluation based on ISO 9001-1987—Purchasing

Reference paragraph number	Function
4.6.1	What form of documentation is used to define the quality-related activities in purchasing goods and services?
4.6.2	How are suppliers selected for the awarding of contracts? *e.g., vendor appraisal visits, quality program approval, qualified supplier lists issued by approval agencies, past performance record, other_____*
4.6.2	How does the purchaser ensure that the vendor quality program remains adequate? *e.g., external quality audits, validation of vendor data, source surveillance, receiving inspection, other_____*
4.6.3	Are the procedures covering the preparation of purchase orders (subcontracts) adequate to ensure: 1. Clear and unambiguous description of the product ordered? 2. The type, class, style, grade, or other precise identification, where applicable? 3. The title or other positive identification and applicable issue of specifications, drawings, etc., including, where applicable, the requirement for approval or qualification of the product, process, operators, etc., to these documents? 4. The title, number, and issue of the relevant quality system standard?
4.6.3	Are purchasing documents verified prior to issuances, and if so, by whom and with what authority?
4.6.3	Are the quality requirements predefined as an aid to purchasing, and if so, how?
4.6.3	Are any "Quality Terms and Conditions" used as an attachment to the purchase order?
4.6.4	Do the procedures allow for the organization's customer carrying out verification activities at a supplier's facility?
4.7	Are procedures prepared, established, and maintained to define the controls applicable to the verification, control, storage, and maintenance activities relating to purchaser-supplied equipment?

APPENDIX 9B.7   Typical Set of Questions for Use in Quality System Evaluation based on ISO 9001-1987—Product Identification and Traceability

Reference paragraph number	Function
4.8	How do the procedures establish and maintain control of the identity of the product and its components during all phases of the program?
4.8	Do the procedures establish a "first-in, first-out" philosophy?
4.8	Do the procedures provide adequate control for the traceability of goods or services where this is required by the contract, regulation, or supporting documentation?

APPENDIX 9B.8   Typical Set of Questions for Use in Quality System Evaluation based on ISO 9001-1987—Process Control

Reference paragraph number	Function
4.9.1	Do the procedures provide adequate process controls, where such processes could adversely affect product quality, during the design and production activities? *e.g., process capability analyses, documented work instructions compatible with the skills and abilities of the user, statistical process (quality) control charts, workmanship standards, facilities, environmental controls, other_____*
4.9.1	Do the procedures provide adequate process controls, where such processes could adversely affect product quality, during the installation, commissioning, and/or decommissioning activities? *e.g., process capability analyses, documented work instructions compatible with the skills and abilities of the user, statistical process (quality) control charts, workmanship standards, facilities, environmental controls, other_____*
4.9.2	Do the procedures provide adequate controls of materials, personnel, process and verification steps involved with special processes such as welding, plating, painting, etc., where such processes could adversely affect product quality, during the design, production, installation, commissioning, and/or decommissioning activities? *e.g., qualification or certification of operator, facility or process, work instructions, quality standards, training program, records, other_____*

APPENDIX 9B.9   Typical Set of Questions for Use in Quality System Evaluation based on ISO 9001-1987—Inspection and Test

Reference paragraph number	Function
4.1.2.2	Does the organization clearly ensure that all inspection and test personnel involved with verification activities are independent of those having direct responsibility for the work being performed?
4.10.1.1	How do the procedures control the flow of incoming goods to ensure none is used or processed until the quality has been verified by some means? *e.g., bonded area, controlled movement of materials, restricted access, receiving inspection, vendor data validation, approval validation, presence of source surveillance, verification by use or application, other_____*
4.10.1.1	Are verification techniques adequately defined for method, sampling, acceptance criteria, etc.?
4.10.1.2	Do some means exist for releasing urgently required goods prior to verification, including identification, means of immediate recall and replacement, controlled application, etc.?
4.10.2	Do the procedures provide adequate controls to ensure: 1. Verification of goods by inspection, test, or other method, in accordance with the quality plan or documented procedures? 2. Establishment of product conformance through the use of process monitoring and process controls? 3. Holding of product, or components thereof, pending verification, subject to controlled urgent usage? 4. Identification and segregation of nonconforming product?
4.10.3	Do the quality plan or procedures require a final confirmation that all verification activities have been completed satisfactorily and the data authorized, as necessary, prior to release of the product for shipment?
4.10.4	Do the procedures clearly define how and when the various inspection and test records are to be prepared, where they are to be held, and for how long?
4.12	How is inspection and test status indicated on the goods and records? *e.g., personalized stamps, traveler cards, labels, verification records, physical location, other_____*
4.12	Where stamps or marks are used to indicate inspection and test status, how are these controlled to prevent unauthorized use?

APPENDIX 9B.10   Typical Set of Questions for Use in Quality System Evaluation based on ISO 9001-1987—Inspection and Test Equipment (Metrology)

Reference paragraph number	Function
4.11	How do the procedures control metrology to ensure control, calibration, and maintenance of all instrumentation, measuring devices, jigs, fixtures, etc., used to demonstrate conformance of the product to the specified requirements?
4.11.a	Do the procedures require the product documentation to clearly identify each measurement required with its relevant accuracy?
4.11.a	How are measuring devices and environment selected to match the accuracy requirements?
4.11.b	How often are measuring devices calibrated, and adjusted if necessary, against standards traceable to the national standards or basic principles?
4.11.c 4.11.d 4.11.e 4.11.f	Do calibration procedures adequately define such factors as periodicity, calibration method and equipment, environment, process control charts (where applicable), marking of calibration status, records, etc.?
4.11.g	Do the procedures require the reassessment of all prior measurements made with a measuring device found to be out of calibration?
4.11.h	Do the procedures control the handling, preservation, storage, and movement of measuring equipment to prevent degradation?

APPENDIX 9B.11   Typical Set of Questions for Use in Quality System Evaluation based on ISO 9001-1987—Nonconforming Product

Reference paragraph number	Function
4.13	How do the procedures ensure that nonconforming product is not inadvertently used or installed? *e.g., identification, documentation, evaluation, segregation, disposition of nonconforming product, notification of functions concerned, other_____*
4.13.1	Is the decision-making chain clearly defined as to responsibilities; disposition, rework, repair, regrading options; customer participation; records, etc.?
4.14	Has the supplier established, documented, and maintained procedures for at least: 1. Investigating nonconformities to determine the cause and the short-term corrective action required on the parts concerned and the longer-term corrective action to prevent recurrence? 2. Analyzing all processes, work operations, concessions, quality records, service reports, customer complaints, etc., to detect and eliminate potential causes of nonconforming product? 3. Initiating preventative actions, where possible, to reduce the risk of nonconformity? 4. Recording all corrective actions taken and the results of the followup activities?
4.13 4.14	Do procedures provide for visibility to management on the extent of the nonconformities detected, the results of the corrective actions, and the outstanding problems associated with nonconformities and corrective action?

APPENDIX 9B.12   Typical Set of Questions for Use in Quality System Evaluation based on ISO 9001-1987—Handling, Storage, Packaging, Delivery

Reference paragraph number	Function
4.15.1	Are there procedures established, documented, and maintained covering the handling, storage, packaging, and delivery of product?
4.15.2	Do the procedures provide for methods and facilities for handling the product that will prevent damage or degradation? *e.g., containers, protective wrap, handling fixtures, loading fixtures, other_____*
4.15.3	Do the procedures provide for storage facilities that will prevent product damage or degradation? *e.g., storage facilities—shelves, bins, containers, etc.; environmental controls; inversion where applicable; segregation; movement controls; other_____*
4.15.4	Do the procedures provide for the design, documentation, verification, and application of packaging methods that will prevent degradation or damage to the product? *e.g., Are design requirements defined? Is the packaging design verified for acceptability? Does the methods data clearly define the packaging and verification method and acceptance criteria? Are periodic verifications required? Other?_____*
4.15.5	Do the procedures provide for determining the most effective mode of delivery, together with any hazards involved that might damage or degrade the product? *e.g., air vs. road vs. sea, environmental hazards, handling facilities at the destination, other_____*

APPENDIX 9B.13   Typical Set of Questions for Use in Quality System Evaluation based on ISO 9001-1987—Quality Audits

Reference paragraph number	Function
4.17	How do the procedures control the scheduling, initiation, planning, implementation, and reporting of internal quality audits?
4.17	Do the internal quality audits evaluate the effectiveness of the activity under review as well as its conformance to procedures?
4.17	Who conducts and interprets the internal quality audits?
4.17	Who receives copies of the quality audit report and any requests for corrective action?
4.17	How are corrective action requests followed up to ensure the proposed action occurs?

APPENDIX 9B.14   Typical Set of Questions for Use in Quality System Evaluation based on ISO 9001-1987—Training

Reference paragraph number	Function
4.18	Does the training program cover quality awareness for all personnel from the executive level down?
4.18	Does the training program cover the development of special skills required for various processes, including the qualification or certification of the personnel where applicable?
4.18	How does the training program cover the requirements for requalification or recertification?
4.18	What records are kept of the results of the training program?

APPENDIX 9B.15   Typical Set of Questions for Use in Quality System Evaluation based on ISO 9001-1987—Servicing

Reference paragraph number	Function
4.19	What procedures are prepared to establish and maintain servicing activities for the product?
4.19	How are the various documents, operating manuals, etc., verified with respect to their accuracy and usability by the intended product users?
4.19	How is the effectiveness of the servicing activity measured and reported?

APPENDIX 9B.16   Typical Set of Questions for Use in Quality System Evaluation based on ISO 9001-1987—Statistical Techniques

Reference paragraph number	Function
4.20	How do the procedures establish and define what, when, and where the various sampling plans are to be used? Which standards cover the sampling plans being used?
4.20	How do the procedures establish and define the types of process capability and process control techniques to be used? How do the procedures establish and define other statistical techniques used? *e.g., design of experiments, cost of quality, Taguchi techniques, analysis of variances, other_____*

APPENDIX 9B.17   Typical Set of Questions for Use in Quality System Evaluation based on ISO 9001-1987—Installation

Reference paragraph number	Function
	What procedures or installation plans cover the installation activities? Is there a separate quality plan or inspection plan developed for the installation program?

Appendix 9C
Typical Working Papers
for Internal Location-Oriented
Quality Audit of a
Manufacturing Facility

APPENDIX 9C.1   Typical Set of Questions for Use in an Internal Location-Oriented Quality Audit of a Manufacturing Facility—Documentation Control

Reference paragraph number	Function
Q1	Check the drawings and specifications at the designated number of production and verification work stations to confirm that: 1. Only correctly released or approved documents are being used. 2. The revision of the drawing or specification being used is the one specified in the work order. 3. The revision history of the document is clearly visible. 4. Copies of drawings and specifications are legible. 5. No unauthorized changes have been made to the documents.

Q2	Check the shop methods data, process instructions, work instructions, etc., at the designated number of production and verification work stations to confirm that:
	1. The instructions are sufficiently definitive to produce the desired results with the skills and facilities available.
	2. The instructions reflect current practice.
	3. The specified tools, facilities, etc. are available and being used in accordance with the instructions.
	4. The specified material, parts, etc., are available and being used.
	5. The personnel and/or processes are certified or approved where required.
	6. The instructions include the necessary verification requirements in a logical sequence.
	7. The verification tools, facilities, etc., are available and being used in accordance with the instructions.
	8. The verification instructions include, as applicable:
	  a. Correct use of approved sampling plan
	  b. Methods of verification compatible with the skills of the operator concerned
	  c. Records to be kept
	  d. Method of showing verification status is defined
	9. The instructions define the necessary safety equipment to be used.

Q3	Check the required number of design change documents per visit to ensure that: 1. The data is complete, legible, and clearly understandable. 2. The proper approvals have been obtained. 3. The effectiveness of the change has been specified.

APPENDIX 9C.2   Typical Set of Questions for Use in an Internal Location-Oriented Quality Audit of a Manufacturing Facility—Materials Handling

Reference paragraph number	Function
Q1	Check at the required number of production and verification work stations to ensure that: 1. Batch controls or traceability controls have been established where required. 2. All material is being handled so as not to degrade it in any way. 3. Handling fixtures, containers, etc., have been provided, along with instructions for use, and are being used correctly to prevent degradation: *e.g., containers, covers, lifting apparatus, other*_____ 4. Movement of materials is being controlled in accordance with the procedures. 5. All material is being properly stored to prevent degradation: *e.g., identification; shelving; inversion, where necessary, of chemicals, paints, etc.; shelf life; containers; other*_____ 6. Environmental protection is provided where necessary to prevent degradation: *e.g., cleanliness, temperature, humidity, vibration, atmospheric pollutants, other*_____ 7. Materials are removed from stores on a first-in, first-out (FIFO) principle. 8. Nonconforming product is clearly identified and segregated from conforming product.

APPENDIX 9C.3   Typical Set of Questions for Use in an Internal Location-Oriented Quality Audit of a Manufacturing Facility—Materials Status

Reference paragraph number	Function
Q1	Check at the required number of production and verification work stations to ensure that: 1. Inspection status of all material is clearly shown: *e.g., stamps, tags, traveler cards, other*_____ 2. Configuration, i.e., drawing or specification number and revision, is identified on all documentation and for the material, goods, etc. 3. Prompt disposition has been made of all nonconforming product or material. 4. Nonconforming material has been properly identified and segregated. 5. Materials and product are clearly identified.

APPENDIX 9C.4   Typical Set of Questions for Use in an Internal Location-Oriented Quality Audit of a Manufacturing Facility—Metrology Controls

Reference paragraph number	Function
Q1	Check at the required number of production and verification work stations to ensure that: 1. Every production tool, gauge, etc., is within the calibration period shown on the item. 2. Only calibrated tools, gauges, etc., are being used for production. 3. Every piece of verification equipment, instrument, tool, gauge, etc., is within the calibration period shown on the item. 4. Calibration records confirm the calibration period for each metrology item used for production or verification that is checked, i.e., gauges, instruments, jigs, fixtures, etc. 5. All metrology items are being used in the environment necessary to achieve the desired accuracy. 6. All metrology items are stored in a manner that will not degrade their performance or accuracy when not in use.

APPENDIX 9C.5   Typical Set of Questions for Use in an Internal Location-Oriented Quality Audit of a Manufacturing Facility—Housekeeping

Reference paragraph number	Function
Q1	Check at the required number of production and verification work stations to ensure that: 1. Each work area is maintained in a clean condition. 2. Each work area is maintained in a safe condition. 3. Aisles and other areas adjacent to the work areas are clear of obstructions, clean, and in a safe condition. 4. Safety equipment is available and used in the defined areas.

APPENDIX 9C.6   Typical Set of Questions for Use in an Internal Location-Oriented Quality Audit of a Manufacturing Facility—Quality Program

Reference paragraph number	Function
Q1	Check at the designated number of production and verification work stations to ensure that: 1. The applicable portions of the quality program procedures, instructions, etc., are available. 2. Every operator is aware of the above and is using them correctly. 3. The applicable standards, acceptance criteria, etc., are available and being used correctly. 4. The applicable identification stamps, markers, etc., are available and being used correctly.

Appendix 9D

Typical Working Papers
for an Internal Quality Audit
of an Administrative Function—
Preparation of a Purchase Order

APPENDIX 9D.1 Typical Set of Instructions for Use in an Internal Conformity Quality Audit of an Administrative Area—Preparation of a Purchase Order

Reference paragraph number	Function
Q1	Check the required number of documents initiating a purchase order or subcontract to ensure that: 1. The initiating documentation in form and usage complies with the applicable procedures: *e.g., requisitions, computer printouts, drawing parts lists, other___*
	2. The initiating documents accurately define the goods or services required: *e.g., part number, drawing or specification number and revision, qualification or certification requirements, description of the goods or services required, quantity, other_____*
	3. All other quality requirements than those already mentioned are clearly defined: *e.g., quality terms and conditions, source surveillance by purchaser, source surveillance by customer, quality system requirements, quality system approvals, vendor data to be provided, other_____*
	4. The supplier has been determined in accordance with procedures: *e.g., vendor appraisal, past performance, qualified product list, other_____*
Q2	Check the required number of purchase orders or subcontracts to ensure that: 1. The necessary approvals to purchase have been obtained. 2. The preparation is accurate and in accordance with the approved procedures. 3. The necessary copies have been prepared. 4. All attachments are in order. 5. Any special notes, stamps, etc., required by the customer are included. 6. The purchase order clearly implies the expectation of error-free performance by the supplier. 7. The purchase order clearly defines the required supplier's actions if a conflicting requirement is noted. 8. The purchase order clearly defines the required supplier's actions if nonconforming material is detected.
Q3	Check the required number of purchase orders to ensure that the work sections responsible for verification of the goods or services being procured are: 1. Aware the procurement has been initiated and know the expected delivery dates 2. Initiating the necessary steps for verification of the goods or services:

APPENDIX 9D.1   Typical Set of Instructions for Use in an Internal Conformity Quality Audit of an Administrative Area—Preparation of a Purchase Order (*Continued*)

Reference paragraph number	Function
	*e.g., vendor quality contact; vendor surveillance; source inspection; receiving inspection; verification of vendor-generated data, i.e., certificates, results, process control charts, etc.; other*_____
	Note: For purposes of this instruction, the following apply:
	1. "Purchaser" denotes the organization buying the goods or services.
	2. "Customer" denotes the organization buying or using the goods or services resulting from the purchaser's activities.

Appendix 9E

Typical Working Papers
for an External Conformity
Quality Audit
of a Service Function—
Airline Travel

APPENDIX 9E.1   Typical Set of Instructions for Use in an External Conformity
Quality Audit of an Airline Traveler Situation—Ticketing

Reference paragraph number	Function
Q1	On entering the ticketing area observe: 1. Is the office clean and tidy? 2. Is the waiting area provided with comfortable chairs, lighting, etc.? 3. Are there publications on traveling available in the waiting area? 4. Are there queues of customers awaiting ticketing? 　a. If so, are the attendants clearly working to clear the queues? 5. Were all ticketing directions clearly visible and understandable?
Q2	While waiting to be ticketed observe: 1. Do all the attendants appear to be giving the passengers their full attention? 2. Did customers in adjacent queues appear satisfied with the handling of their business?
Q3	While purchasing your ticket observe: 1. Did the attendant ensure that your flight times and connections were the most suitable for you? 2. Did the attendant offer any fare- or time-saving alternatives? 3. Was your ticket handled in accordance with the defined procedures? 4. Was the attendant courteous and helpful? 5. Were flight times and flight transfers clearly explained and visible on the ticket? 6. Were the times for "flight book-in" clearly stated verbally and in writing? 7. Were the luggage and carry-on rules clearly explained?

APPENDIX 9E.2   Typical Set of Instructions for Use in an External Conformity Quality Audit of an Airline Traveler Situation—Departure

Reference paragraph number	Function
Q1	Was the departure and flight information clearly displayed?
Q2	Were the check-in directions clearly visible and understandable?
Q3	Were attendants in your line and adjacent lines handling their tasks courteously and in an organized manner?
Q4	Was your boarding pass correctly issued?
Q5	Was the boarding and security clearance information clearly stated and understandable?
Q6	Was your luggage handled properly?
Q7	Were passengers in need of luggage or other assistance or help receiving it?
Q8	Was carry-on luggage handled correctly and courteously by the security personnel?
Q9	Was the boarding area clean and tidy?
Q10	Was the boarding gate clearly posted and understandable?
Q11	Were boarding instructions clear and understandable?
Q12	Were boarding instructions handled in accordance with procedures?
Q13	Was boarding handled in a logical sequence and according to the guidelines?
Q14	Was the security check at the boarding effective and the same for all passengers?

APPENDIX 9E.3   Typical Set of Instructions for Use in an External Conformity Quality Audit of an Airline Traveler Situation—On-Plane

Reference paragraph number	Function
Q1	Were seating directions clear and understandable?
Q2	Were the attendants courteous and helpful on boarding?
Q3	Was boarding carried out in a way that permitted the aisles to be kept clear?
Q4	Were the safety instructions clearly presented and understandable?
Q5	Were the airline magazine, safety instructions, etc., in the seat pocket?
Q6	Had all extraneous material, garbage, etc., been cleared from your area and the aisles?
Q7	Were the seats, head rests, cushions, etc., clean and protected, as required?
Q8	Did the cabin crew ensure that all carry-on luggage, shopping bags, etc., were stowed safely away during takeoff, the flight itself, and landing?
Q9	Were washrooms clean and properly stocked?
Q10	Were refreshments, food, etc., handled courteously and correctly? 1. Was the hot food hot and the cold food cold? 2. Were refreshments satisfactory? 3. Were all refreshments tastefully served in a pleasing manner?
Q11	Was in-flight service satisfactory, i.e., prompt, courteous, etc.?
Q12	If it was an international flight, were the necessary customs and immigration forms provided in time for completion prior to landing?
Q13	Were deplaning instructions clear and understandable?

APPENDIX 9E.4   Typical Set of Instructions for Use in an External Conformity Quality Audit of an Airline Traveler Situation—Arrival

Reference paragraph number	Function
Q1	Were the ground attendants handling special assignments clearly identified?
Q2	Were luggage claim instructions clear?
Q3	If applicable, were customs and immigration instructions clear?
Q4	Was luggage available at the claim area in a reasonable time and easy to recover?
Q5	Was any check made to determine if you had claimed your own, and only your own, luggage?
Q6	Were exit instructions clear and understandable?

10

Quality Audit Sampling Plans and Decision Sampling

10.1 Introduction

There are two basic truths that should be stated at the start of this chapter:

1. All sampling plans used during the quality audit must have sound statistical foundations.
2. All activities involving numerical analyses must have sound mathematical foundations.

This chapter applies to quality audits used to determine conformity to quality program documentation, i.e., to procedures, instructions, etc. It does not apply to quality audits performed to determine the suitability of that documentation with respect to some reference standard. Suitability quality audits require a point-by-point comparison between the reference standard and the documentation forming the quality program to be audited. Hence, no form of statistical sampling applies.

Conformity quality audits are instituted to determine if the actual work practices agree with the procedures, instructions, etc., defining the quality system. In virtually every situation likely to arise, the number of actions, decisions, etc., will be too large to evaluate each individually. Thus an auditor is always in the position of having to make a decision based on a sampling of the population affected by what is being audited.

Fundamentally an auditor is evaluating a situation where the activity is either satisfactory or not satisfactory. Thus, we are dealing with attributes sampling rather than variables sampling. If errors in decisions or failures to conform to the desired procedures, etc., are present, they will be expressed in complete integers. Therefore, we are dealing with discrete probability distributions. The most common of these used in sampling plans are hypergeometric, binomial, and Poisson.

Normally an auditor will not know the size of the population being evaluated. However, in most cases, the number of incidents having a potential error or nonconformity will be large. A rare exception to this could be metrology, where the controls should involve a known inventory of devices.

Therefore, a sampling plan will normally be based on a distribution where the population is large but unknown and the probability of acceptance is dependent on the sample size only. The binomial and Poisson distributions satisfy this requirement with no qualifying hypotheses. The hypergeometric distribution involves a finite population, which becomes a factor in the probability calculations. The impact of the population size (N) becomes negligible when it is much greater than the sample size (n), i.e., when $N \geq 10n$.

Any detection of error or nonconformity in the sample indicates a need for corrective action. Thus we are looking for error-free performance and hence a zero acceptance number. Therefore, the sampling plans should reflect this.

Traditional zero-acceptance sampling plans are not sufficiently discriminating between acceptable and unacceptable lots, largely because they were developed for use with acceptable quality levels (AQLs). Limiting quality level (LQL) sampling plans, with the selected values for the LQL and probabilities of acceptance being related to the expected performance and the risks involved in quality audit sampling, provide the necessary discrimination.

The limiting quality level is defined in ANSI/ASQC A2-1987 as follows:

> The percentage or proportion of *variant units* in a batch or lot for which, for the purposes of acceptance sampling, the consumer wishes the probability of acceptance to be restricted to a specified low value.

LQL has also been, and is still, known by a variety of other names including rejectable quality level (RQL), unacceptable quality level (UQL), limiting quality (LQ), and, when the percentage of nonconforming units is used as the measure, lot tolerance percent defective (LTPD).

Certain published standards relating to LQL and LTPD may be of value in those situations where the population has a finite size. In most cases, however, these standards will have error rates and probabilities of acceptance that are too high for good auditing practices.

10.2 Decision Sampling

Decision sampling is a special application of zero-acceptance sampling plans that evaluates the ability of decision-making individuals to make valid quality decisions. Practice has shown that this technique is particularly helpful in evaluating verification decisions, although it can also be used to evaluate many other activities, e.g., auditing, supervisory, design, etc.

It is my hypothesis that an individual engaged in verification activities, i.e., an inspector, tester, chemical analyst, drawing checker or approver, proofreader, etc., is fundamentally paid to make decisions, even if some of those decisions are based on the results of measurements and observations. The nature of the measurements or observations and being able to understand their import may determine the skill level of the individual making a decision based on them; however, dealing with such data is still only a means to an end—that of making a decision. In the following paragraphs the term "inspector" applies to any individual making such decisions in order to carry out a verification activity, regardless of the actual job title or whether it involves working with measurements and observations.

A valid quality decision can take one of two forms: the quality level is "acceptable" or "not acceptable." Therefore, the samples evaluated must be taken from the results of both types of decisions. In this way, decision sampling provides:

1. Protection to the customer against acceptance of nonconforming product, goods, or services
2. Protection to the producer against the nonacceptance of conforming product, goods, or services
3. Information on shortcomings in the decision process

All sampling plans include the risk of making a wrong decision. Therefore, it is important that the risk incurred by an auditor making a wrong decision be known and allowed for in selecting the decision sampling plan to be used. However, where verification is based on sampling, while the risks incurred by incorrect decisions being made in relation to any individual item being measured must be a factor in the ultimate quality evaluation, such risks have no bearing on the determination of the decision sampling plan that should be used.

Correct decisions, either to accept or reject, by an inspector pose no problems to users. Correct "reject" decisions do create the problem of remedial and preventive corrective action. However, this corrective action can be based on fact.

On the other hand, incorrect decisions both to accept and to reject can create problems for both users and the producer. Incorrect acceptance decisions result in:

1. Faulty product getting into the hands of users
2. A reduced reputation for quality goods, with a potential loss of sales, for the producer

Incorrect rejection decisions result in:

1. Additional costs and delays due to reworking acceptable product
2. Wasted problem solving and corrective action effort, resulting in increased costs
3. Possible loss of business due to increased costs and delays

When used as a tool for an internal quality audit, decision sampling provides a measure of the consistency of the decisions being made. The data derived lends itself to the development of process control charts for individuals to determine work performance or trends, and when the data for individuals is combined, to the development of a control chart for a group of individuals working at the same basic occupation, which provides a means of comparing an individual's performance to that of the group. Based on the data obtained, training programs, facility or methods changes, etc., can be introduced to improve the decision-making process by correcting identified weaknesses or shortcomings.

The principles discussed apply to both external and internal quality audits. However the greatest value is experienced with internal quality audits owing to their repetitive nature and the opportunity to monitor trends.

10.3 Sampling Plan Definitions

There are four key terms in quality audit sampling (not currently defined in any standards in the form given here):

1. *Acceptable performance level (APL).*   The acceptable performance level is the lowest performance level that can be considered as acceptable for the function being audited. Since the performance objective should be 100 percent error-free performance, the APL does

not represent a performance objective. It is used solely to determine sample sizes. It is the LQL value for the process.

2. *Acceptable error rate (AER).*   The acceptable error rate is the maximum error rate that can be considered as acceptable for the function being audited. It does not represent an error rate objective. The objective for all processes, decision-making activities, etc., should be error-free performance, i.e., zero defects. AER is used only in determining the relevant sample size. It forms the LQL, expressed as an error rate, for the process.

Note: APL and AER are the converse of each other, i.e., APL + AER = 1.

3. *Confidence level (C).*   The confidence level expresses the degree of certainty that the selected sample contains at least one example of any errors that are present.

4. *Risk (R).*   Risk expresses the degree of uncertainty that the selected sample contains at least one example of any errors that are present.

Note: C and R are the converse of each other, i.e., $C + R = 1$.

10.4   Binomial Sampling Plan

10.4.1   Binomial expansion

As stated above, a quality auditor works in a binary mode, i.e., the decision is selected from two alternatives, OK or NOK, acceptable or unacceptable.

Binomial sampling plans are based on the binomial expansion:

$$(q + p)^n = (p + q)^n = (\text{AER} + \text{APL})^n = 1 \qquad (10.1)$$

where q = decimal quality level in the population
$\quad p$ = decimal nonconformity level in the population
$\quad n$ = sample size
$\quad \text{AER}$ = acceptable error rate = p
$\quad \text{APL}$ = acceptable performance level = q

Expanding Equation (10.1) gives:

$$q^n + \cdots + p^n = 1$$
$$P_0 + P_1 + \cdots + P_{(n-1)} + P_n = 1$$

where $\qquad P_0$ = probability of finding zero nonconformities in a sample size of n and a quality level of q

P_n = probability of finding n nonconformities in a sample size of n and a quality level of q or $1 - p$

$P_1, P_2, ..., P_{n-2}, P_{n-1}$ = probabilities of finding 1, 2,..., $n - 2$, and $n - 1$ nonconformities in a sample size of n and a quality level of q

10.4.2 Determining sample size

Since in binomial sampling we are interested in determining the probability that the sample will include at least one example of any errors present, the confidence that this is achieved is the sum of all the probabilities of the sample size including at least one example of any errors present, i.e., $P_1 + P_2 + ... + P_{(n-2)} + P_{(n-1)} + P_n$.

Conversely the risk of not achieving this is represented by P_0. But

$$P_0 = q^n = (1 - p)^n \qquad (10.2)$$

Solving for n:

$$\log_{10} P_0 = n(\log_{10} q) = n[\log_{10}(1 - p)]$$

Transposing:

$$n = \frac{\log_{10} P_0}{\log_{10} q} = \frac{\log_{10} P_0}{[\log_{10}(1 - p)]} \qquad (10.3)$$

where P_0 = risk (R) of not detecting at least one example of any errors present, when using a sample size n

q = decimal quality level of the population, i.e., the acceptable performance level

p = decimal nonconformity level of the population, i.e., the acceptable error rate

In addition:

C = confidence level that the sample size will detect at least one example of any errors present

$= 1 - P_0$

Thus, using either form of Equation (10.3), the sample size n can be determined for various values of either APL or AER and with selected values of confidence levels C or risks R. In most applications, the calculations will normally be based on low values of R, e.g., 0.05 or less, and low values of the AER, e.g., 0.05 or less. This implies a minimum

TABLE 10.1   Binomial Limiting Quality Level Sampling Plan for a Quality Audit

Acceptable performance level	Confidence levels					
	0.950	0.97	0.98	0.99	0.995	0.999
	Sample sizes					
0.90	28	33	37	43	50	66
0.95	58	68	76	90	103	135
0.96	73	86	96	113	130	169
0.97	98	118	128	151	174	227
0.98	148	174	194	228	262	342
0.985	198	232	259	305	351	457
0.99	298	347	389	458	527	687
0.993	426	499	557	656	754	983
0.995	598	700	780	919	1057	1379
0.997	997	1167	1302	1513	1763	2244
0.998	1496	1752	1954	2300	2647	3450
0.999	2994	3505	3910	4603	5296	6904

sample size of 58. A binomial LQL sampling plan based on these conditions is given in Table 10.1.

10.4.3 Application

Since the population concerned is the collection of decisions made by one or more individuals, the sample size can be spread across the typical decisions within a particular group. It may also be spread over a period of time, providing the selection is still made on a random basis.

Example Consider Q1 of Appendix 9C.1. This question involves five topics to be evaluated with respect to drawings and the same number with respect to specifications. The evaluation is being carried out at a single location. However, let us assume there are differences in the control cycles for drawings and specifications. It has been determined that the acceptable performance level for drawings is 0.95 and for specifications is 0.97. A confidence level of 0.95 is required in each case.

From Table 10.1 the sample size for drawings is 58 and for specifications is 98. For drawings:

$$n = 58$$

Number of activities to be evaluated = 5

$$\frac{n}{\text{Number of activities}} = \frac{58}{5} = 11.6$$

Since it is impossible to select 0.6 samples, the number of evaluations per activity will be 12. Therefore, the evaluation of the decisions on drawings will be based on a total sample size of 60.

For specifications:

$$n = 98$$

$$\frac{n}{\text{Number of activites}} = \frac{98}{5} = 19.6$$

or 20 samples per factor.

Therefore, the evaluation of the decisions on specifications will be based on a total sample size of 100.

conclusion: If no nonconformities are found in either case, it can be stated with 95 percent confidence that the performance level on drawing controls is at least 0.95 and for specification controls at least 0.97.

10.5   Poisson Distribution

10.5.1   Principles

The Poisson distribution is used to determine the probability of the occurrence of a specific number of incidents in a given population n where the average number of incidents (p_0) over a series of populations is known. Once p_0 is known, the average number of incidents in the sample size n is the product np_0. From this, the probability of different numbers of incidents occurring in a sample size n can be determined. The Poisson distribution formula is:

$$P_c = \frac{X^c}{C!} e^{-x} \qquad (10.4)$$

where $x = np_0$

   n = sample size

   p_0 = average error rate

   np_0 = number of incidents expected in a sample size n

   c = the actual number of errors, in the sample, to be considered

   P_c = probability of c nonconformities being found in the sample of n items

In developing a zero-acceptance-number sampling plan, $c = 0$. Therefore, from Equation (10.4):

$$\frac{X^c}{C!} = 1$$

Therefore

$$P_0 = e^{-x}$$

Transposing to solve for x:

$$x = \log_e P_0 = np_0 \qquad (10.5)$$

TABLE 10.2   Table of np Values for Selected Probabilities of Acceptance

Probability of acceptance	0.10	0.05	0.03	0.02	0.01	0.005	0.001
np_0	2.303	2.996	3.507	3.912	4.605	5.298	6.908

From Equation (10.5), np_0 values can be calculated for various probabilities of acceptance. In a quality audit, we are looking for high levels of probability that the selected sample will include at least one example of any errors present in the area being evaluated. Therefore, the probabilities of acceptance must be low. Substituting for various values of the acceptance probabilities, Table 10.2 can be derived.

10.5.2   Determining the sample size

Having derived the np_0 values, the sample size can then be calculated for various values of p_0, which represents the acceptable error rate. These calculations result in Table 10.3, which relates sample size to the desired acceptable error rate.

10.5.3   Application

From Table 10.3, the sample size can be selected for the appropriate values of AER and probability of acceptance (P_a). Since the decisions can be spread over a wide range of factors, the samples may be divided among the various factors involved, depending on the criticality of the parameters concerned. This grouping must

TABLE 10.3   Poisson Limiting Quality Level Sampling Plan for a Quality Audit

Acceptable error rate	Probability of acceptance						
	0.1	0.05	0.03	0.02	0.01	0.005	0.001
	Sample size						
0.10	23	30	35	39	46	53	69
0.05	46	60	70	78	92	106	138
0.04	58	75	88	98	115	132	173
0.03	77	100	117	130	154	177	230
0.02	115	150	175	196	230	265	345
0.015	153	200	234	261	307	353	461
0.01	230	300	350	391	461	530	691
0.007	329	428	501	559	658	757	987
0.005	461	599	701	782	921	1059	1382
0.003	768	999	1169	1304	1535	1766	2303
0.002	1151	1498	1753	1956	2303	2649	3454
0.001	2302	2996	3507	3912	4605	5298	6908

be watched carefully as the number of samples in a single factor should, as a rule of thumb, be five or more.

Example   If we consider the "housekeeping" activities described in Appendix 9C.5, we find there are four factors to be checked. If it is desired to use a probability of acceptance of 0.01 and an acceptable error rate of 0.03, from Table 10.3 it is found that a sample size of 154 is required. As there are four factors to be checked, this results in 39 samples per factor to be checked.

Since one of the factors is concerned with safety, this could arbitrarily be given a weighted value of five times the other factors. This approach would effectively result in eight factors. With this approach, we have 20 samples per factor other than safety, along with 100 safety checks.

conclusion: If zero nonconformities or departures from the defined methods and techniques are found, it can be stated that: "There is only a 1 percent risk that the performance in this area has an error rate in excess of 3 percent."

10.5.4   Graphical presentation

Table 10.2 provides the np_0 values from which a graphical presentation of Table 10.3 can be developed relating the values of sample size n to the acceptable error rate p_0. This will be a series of straight-line diagonals on log/log graph paper, where each diagonal represents a probability of acceptance P_a. Single-cycle log/log paper can be used if the reader is thoroughly familiar with this type of presentation but multicycle papers provides a clearer presentation, particularly if the user has difficulty converting the decimal points.

For those not used to log/log graph paper, "single cycle" means an axis is scaled with a single decade of values from 1 to 10; "multi-cycle" means there are more than one decade of values along the axis, with the plotter normally determining the positions of the decimal points. Graph paper can be purchased with differing numbers of cycles along the horizontal and vertical axes. In plotting Table 10.2, two cycles would cover the selected values of AER shown in Table 10.3 (three cycles would allow the extending of these values); similarly, three cycles would be required to provide coverage of sample sizes from 10 to 10,000. Therefore, three-cycle by three-cycle log/log graph paper should be used.

10.6   Published Limiting Quality Level Sampling Plans

10.6.1   Introduction

This section will examine some of the published standards that relate to limiting quality level sampling plans under its various titles.

10.6.2   Sampling plans

LQL sampling plans are all attribute sampling plans. Most published

LQL sampling plans are based on the use of finite lot sizes or populations and thus have limited application in quality auditing. One exception is MIL-S-19500, which includes an LTPD sampling plan where the sample sizes are related to the desired acceptance number and LTPD.

Probably the best known attribute sampling standard containing LQL information is MIL-STD-105, now updated and reissued as ANSI/ASQC Z1.4-1981, *Sampling Procedures and Tables for Inspection by Attributes*. Tables VI-A and VII-A of the ANSI/ASQC standard provide LQL values as the percentage of nonconformities at P_a = 0.10 and 0.05, respectively. These values are given for the normal acceptable quality level sampling plans. Hence the sample sizes are relatively small, for zero acceptance. This results in an operating characteristic (O/C) that does not provide adequate discrimination. Similarly the o/c curves are too cramped in the areas of concern to be of value.

MIL-S-19500 also has limited application in the quality audit field. Some of the LTPD levels (average error rates) it gives are sufficiently low to be of value. However, the sampling plan is based on a single probability of acceptance P_a = 0.10, which may be too high for many applications. However, the standard does include tables showing the LTPD values of given sample sizes for lots or populations smaller than 200. This could be of value when auditing an activity with small but finite lot sizes.

A new standard, ANSI/ASQC Q3, titled *Sampling Procedures and Tables for Inspection of Isolated Lots by Attributes*, is under development. This will be a limiting quality level sampling plan document. As work on it is not yet complete, it is too early to be certain of its application to quality audits, but initial indications are that it may have a limited usefulness.

10.7 Application of LQL Sampling Plans

In looking at the use of the binomial and Poisson limited quality level sampling plans discussed above, there are certain points to be borne in mind:

1. Conformity quality audits fundamentally evaluate the ability of individuals to follow the procedures, instructions, etc., defining a quality program.

2. In virtually every case, the individual is making a decision of some type with respect to the quality of the operation or activity concerned.

3. Thus the homogeneity of the sample consists of the decisions made, albeit those decisions may relate to different technical points.

4. Since decisions are being made continuously in any given area

based on a mix of elements, grouping can be done by combining elements per visit or elements per period of time, or even by individuals, to determine the performance of a group.

5. This flexibility is of greatest value to internal quality auditors where it can be used to determine trends through the use of process control charts, etc.

6. When grouping is done by period of time, care must be taken to select the samples at random times over that period. These can be stratified by taking so many per subperiod of time and then using industrial engineering random time tables to take the actual sample.

10.8 Conclusions

From the definition of "quality audit" and the above discussion, it becomes readily apparent that sampling in a quality audit situation is basically a sampling of the acts and decisions made by various individuals functioning within the quality system. There will be a large number of these decisions for each individual, spread over a variety of factors.

Since the common factor is decision making itself, many factors can be grouped to form the homogeneous sample. It is recommended that only those decisions having similar acceptable error rates and carrying the same risks be grouped together. However, there will be occasions when the grouping may involve weighting factors to allow for differences in the AER.

The above discussions show that quality audits must use some form of limiting quality level sampling plan. However, none of the currently published standards that I am aware of provide a suitable range of LQLs and probabilities of acceptance (P_a).

Either of the sampling plans summarized in Tables 10.1 and 10.3 can be used, with little to choose between them. Psychologically speaking, the binomial sampling plan has the advantage of being well known as a means of calculating probabilities in a two-state situation, whereas the Poisson does not have the same general appeal. But in both if values outside those shown are required, they can be readily calculated.

11

Implementation of the Quality Audit

11.1 Introduction

There are many similarities between the implementation of an external quality audit and of an internal quality audit. This is evident by an examination of Figures 11.1 and 11.2.

However, there are sufficient differences in approaches and techniques to justify examining each in turn. The authors of the ASQC and CSA standards on quality audits have recognized the similarities and the differences and by judicious selection of words and phrases have made these standards applicable to both types of audit.

First we will consider the implementation of an external quality audit, i.e., the key points in the third level of such audits, as shown in Figure 11.1. Then we will turn to the implementation of an internal quality audit, again considering each of the key points in the third level of such audits, as illustrated in Figure 11.2.

11.2 External Quality Audits

11.2.1 Opening management meeting

11.2.1.1 Purpose. The initial contact between the auditing organization, the client, and the auditee will have been made at the time of initiating the quality audit, as discussed in Chapter 5.

The opening management meeting should be attended by:

1. Executive officers of the auditee, or if the facility is a subunit of a larger corporation, the upper levels of management for the facility under review. Among these must be the individual responsible for

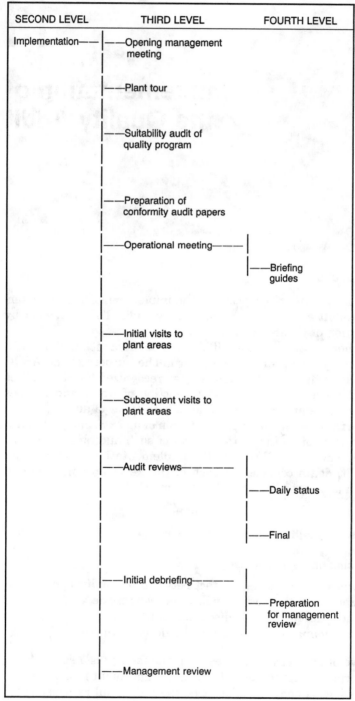

Figure 11.1   Function tree for the implementation of an external quality audit.

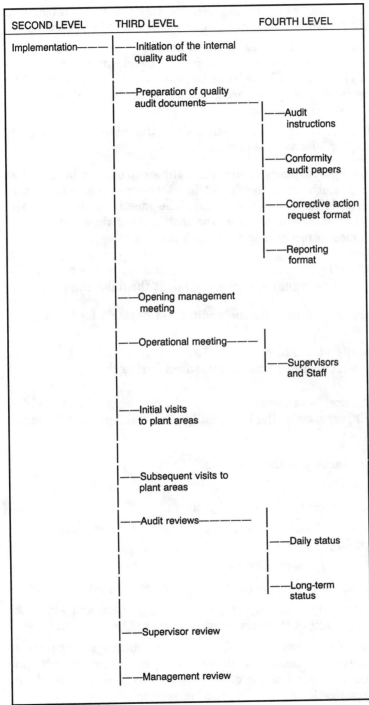

Figure 11.2 Function tree for the implementation of an internal quality audit.

resolving quality problems and coordinating the audit on behalf of the auditee. This individual can carry a myriad of different titles.

2. The members of the auditing team, including, where possible, the supervisor of the auditors. The arrival of the quality auditors at the auditee's facility should be timed to coincide with exact start of this meeting so they have no direct contact with the plant until after completion of the meeting.

3. The client or a senior representative of the client's organization. See Chapter 3 for typical clients.

Each of these participants have certain things to offer or bring out at the meeting. Each attendee should leave the meeting with certain types of knowledge or understanding. The meeting can most effectively be chaired by the client as the instigator of the audit and the individual most directly concerned with its results.

11.2.1.2 Client involvement. The client must ensure that a *mutually-agreed-upon* understanding is reached about the following points:

1. The purpose or reason for the audit. See Chapter 3 for typical purposes.

2. The quality standard that is to form the reference or benchmark against which performance is measured during the audit.

11.2.1.3 Auditor involvement. The auditing organization, either through a supervisor or the lead auditor, must ensure that there is a *consensus* on:

1. The acceptability of the members of the audit team

2. The method of conducting the audit

3. The points of contact in case of problems

4. The degree of detail to be provided at the final management review or debriefing

5. The schedule of audit operations

6. The assistance or support required by the auditors from the auditee

7. The availability of quality program documentation and where the suitability audit of this documentation should be conducted

8. The degree of commitment of the auditee's management team to the quality program, and their awareness of its requirements and the need for control of the quality system as it is integrated into their organization's overall operating system

In developing the consensus on these issues, the auditing organization may have to provide such information as:

1. Evidence to support the competence, independence, and objectivity of the auditors.

2. Working papers to indicate the method of conducting the audit. In this connection, I see no problem in providing the auditee's management team and the client with a set of working papers showing the various items to be reviewed in evaluating the compliance of the auditee's procedures or product with the reference standard. This can serve to demonstrate that there is no hidden agenda for this comparison. To prevent confusion, it is probably best to provide copies without the performance levels or sample sizes entered.

3. A schedule for the various activities to be conducted.

11.2.1.4 Auditee involvement. The auditee must help achieve a *consensus* on the following points:

1. The individual who will represent the auditee on all matters during the audit

2. Auditor access to the various areas and activities to be audited

3. The facilities to be provided for use by the auditors in the auditee's plant

4. The support personnel who will be provided during the audit, e.g., guides, specialists, line personnel, etc.

5. How safety and other regulatory controls will be met during the audit

6. How the organization's proprietary rights will be protected in the course of the audit

In order to achieve consensus or mutual agreement on those points, the auditee should bring to the meeting:

1. All documentation relating to the organization's quality system, e.g., copies of the quality policies, quality manual, quality program procedures, etc.

2. Evidence of management support of and commitment to the quality system

3. Information on the availability of facilities and people to assist the auditors, where requested, recognizing that the time these individuals may have for audit purposes might be limited owing to other commitments

11.2.1.5 Conclusions. If the attendees or their assistants have done their homework in preparing for the meeting, the meeting itself should take no longer than an hour. This includes the time necessary to evaluate the auditee management's commitment to and support for the quality system.

Once the meeting is completed, the members of the auditee's management team must brief subordinates on the audit, how it will effect different work areas, and the inputs expected from employees. This briefing is essential to ensure that first-level supervisors are aware that people will be visiting the work areas for which they are responsible, examining processes, products, facilities, etc., and asking questions of personnel.

11.2.2 Plant tour

A fairly quick plant tour after the meeting can provide the auditors with a feel for the facility, its general layout, working environment, work tempo, flow of goods, product, etc. The most effective route for this walk-through is to follow the organization's flow of ideas, documents, goods, and materials, i.e., to move from marketing, through design, materials control, production, packing, shipping, and servicing. In a production area, the tour should follow the flow of goods and materials from the receiving area through shipping of the completed product.

During the initial tour, no attempt should be made to delve into how things are handled or controlled; rather auditors should concentrate on the sequence and locations of the various activities. If an auditor observes something questionable or amiss, a mental note should be made for later verification during the audit. Of course, if something serious is noticed, it should be questioned on the spot and resolved.

11.2.3 Suitability audit of the quality program

The auditors must have copies of the reference standard and the applicable working papers available, in sufficient numbers to cover any activities appearing in more than one functional area (see Figure 9.5).

Once the quality program documents have been brought together, the coordinator for the auditee should run through them with the auditors. This will establish such things as order of precedence, who uses particular documents, and where copies are available in the plant.

A suitability audit is a point-by-point checking of various activities or work areas against the reference standard, both from a general point of view and in relation to details of application. Normally during this audit, only those points covered by the requirements are reviewed in detail—for example, an organization may design certain products, but if it is being evaluated for its ability to produce goods under the ISO 9003-1987 standard, which makes no mention of design and only requires final inspection, the design elements need not be audited in detail. However, if the review is part of a vendor appraisal, the additional points should be considered since they could be of added value on a given contract or reveal a source for future business.

As each activity or area is reviewed, a decision must be made as to whether it meets the standard's requirements or not. This decision may be of the simple OK/NOK type or it may be a somewhat more helpful grading of satisfactory, adequate, inadequate, or unsatisfactory. The decision should be recorded on the working paper for later analysis as part of the total performance picture.

A form such as that shown in Figure 9.5 becomes an excellent way of summarizing the state of the documentation. Different colors or codings can be used to indicate what decisions were made.

11.2.4 Preparation of conformity audit papers

As the suitability quality audit is being conducted, notes should be made of the detailed points to be verified during visits to the various working areas. These notes are best prepared on blank working paper sheets. Each working area to be evaluated should have its own working paper, along the lines shown in Apendixes 9C and 9D. The confidence level, or conversely the risk level, the performance level (APL or AER), and the sample size per question should be defined as part of the audit instructions for the particular standard, in order to maintain continuity from audit to audit.

In preparing the detailed instructions, it must be remembered that in most organizations many of the quality program elements apply to a variety of functions (see Figure 9.5). For example, in a manufacturing organization, documentation controls will apply to virtually every function being evaluated. Thus, by dividing the documentation sample over the various activities, a fairly complete picture can be obtained of the effectiveness of these controls. Similarly, metrology controls will apply to virtually every area except marketing and purchasing. (Groupings of this nature are discussed in Chapter 10.) The sample size per activity or work area being audited should be entered on the working paper for that activity or area. Information about such multifunction quality system elements is used only to verify that system procedures are being followed throughout the organization; it does not replace the basic questions to be used in the various areas.

11.2.5 Operational meeting

This meeting consists of the audit team and the auditee's quality authority and those members of the auditee's staff who will be acting as guides throughout the plant (normally the latter will be members of the quality function). At this meeting, *mutual agreement* should be reached on:

1. The sequence and timings of the visits to areas within the organization, recognizing that if the audit is conducted by an individual the areas will be visited in series, while if it conducted by a team they can be evaluated in parallel.

2. The matching of auditors with guides, according to the skill and knowledge demands of the areas to be audited.

3. The routine and protocol to be followed during the visits to the various areas. For example, who speaks for the quality system in a given area? Is it the quality coordinator, the guide as the coordinator's delegate, the area supervisor, or the individuals carrying out tasks in the area, i.e., the individual designer, operator, inspector, tester, etc.?

11.2.6 Initial visits to plant areas

When an auditor and guide first arrive in an area of the plant, the first step is to introduce the auditor to the supervisor responsible for the functions involved. The supervisor may be in charge of a productive function, the related assurance function, or both. Despite the limitations placed on organizations by some procurement quality standards, one will frequently find that a supervisor is both responsible for the work done in a department and accountable for the quality of that work, i.e., in charge of both the productive and the verification activities of a department—for doing the work properly and demonstrating that the obligation to do it properly has been fulfilled. Personally, I think this situation is the most suitable for carrying out an appropriate internal quality audit activity.

In any of the major work functions, i.e., marketing, design, etc., the inputs and responses of the productive function to the quality system should be covered first. Then the verification activities should be evaluated. The supervisor should be asked to describe the activities in the area: how they are initiated, defined, controlled, and verified. These descriptions should provide many answers to the questions on the working papers. Where more information is needed, polite questioning should suffice to elicit it.

As the various points are covered, the supervisor should be requested to show examples and note whether these agree with descriptions of the activities and appear to be typical. In cases like drawing control, some of the drawing numbers and revisions should be noted for checking against the work order and also against the configuration control information in the drawing office. Similarly, as measuring equipment is checked, its type, serial number, and calibration status, if marked, should be recorded. This data can then be checked against the metrology records to see if the instrument data is compatible with the controls required and if the equipment is in the right location. These kinds of notes for cross-checking controls are effective and provide audit evidence regarding both the controls in general and the details of controls within a given location or function.

As the definition of a quality audit implies, the examination and evaluation of a work area or activity must cover people, facilities, product, and records. Therefore, while touring a location with the supervisor, an auditor should use all the senses to locate clues about everything happening in that location. Relying totally on the supervisor or guide for information can easily lead to missing key points in the audit. Unless previously agreed on as an unacceptable procedure, the auditor should query some of the individual operators carrying out the work activities to determine their awareness of the quality system, their perceptions of management's commitment to it, and their understanding of how it effects their activity.

11.2.7   Subsequent visits to plant areas

During an audit, it is sometimes necessary for an auditor to return to a location to verify or clarify points noted during the initial visit. In these cases, the auditor and guide should not enter a work area without letting the supervisor know they are doing so. However, the supervisor should not be requested to be present during these additional visits unless it is absolutely necessary. This will help reduce interference with the normally busy schedule of an individual who is always working under conflicting pressures.

11.2.8   Reviews

11.2.8.1   Audit team reviews.   When the quality audit is being conducted by a team, team members should meet at the end of each day to review their findings. Out of this may come changes in the plan for the next day if some additional checking is required to clarify outstanding points. The findings by members in one area may also result in warnings to other members of the team about potential problems and possibly suggest additional activities for the auditors researching other related areas.

Once the audit has been completed, the team should review its findings and develop a preliminary report. This preliminary report should, if possible, be in quantitative terms and deal with specifics. If the analysis cannot be completed immediately, this should be noted in the preliminary report. Writing up reports will be covered in more detail in Chapters 12 and 14, "Analysis of the Quality Audit" and "Reporting the Quality Audit," respectively.

11.2.8.2 Reviews with auditee's coordinator. Each morning the previous day's findings should be reviewed briefly with the auditee's coordinator. This meeting should be used to clarify or correct any misunderstandings that may have arisen during the previous day's activities. These clarifications or corrections should in turn be checked to ensure the revised interpretations are correct.

This meeting can also be used to announce any changes in the current day's planned activities resulting from requests by either the auditee or the auditors.

11.2.9   Initial Debriefing

When the analysis at the end of an audit has been completed and the preliminary report has been prepared, it should be reviewed with the auditee's quality coordinator. In this way, the coordinator will be aware of any shortcomings before the management debriefing is held and will not be subjected to unpleasant surprises. If the report offers bouquets it might be desirable to hold these for the management review, in order to increase their impact.

11.2.10   Management Debriefing

At the conclusion of the audit and before the audit team departs, a preliminary review or debriefing should be held with the attendees of the original meeting. The report provided to this meeting is dealt with more fully in Chapter 14. At this point suffice it to state that the preliminary report should include:

1. An appreciation of the cooperation and assistance provided by the auditee's management and staff. If there have been some special points or areas of help, these should be highlighted.

2. A general statement on the acceptability of the quality system. It may be necessary to stress that the final decision will be made in writing once the detailed analyses have been completed.

3. A summation of the strong points about the quality system noted during the audit.

4. A summation of the major areas requiring corrective action. Care must be taken to ensure that all major shortcomings are mentioned, so that the written report will not introduce any surprises in the way of major new problem areas not previously noted.

5. A summation of the minor areas requiring corrective action.

6. A summation of the quality audit activities remaining to be carried out, e.g., the final written report, the methods to be used in handling corrective actions, the possible followup visits to confirm that the corrective actions have been taken, etc.

11.3   Internal Quality Audit

11.3.1   Introduction

The general principles that apply to implementing an external quality audit apply equally well to implementing an internal quality audit. This similarity reflects on the thoroughness of the work done by the writing committees of ASQC and CSA in preparing their standards on quality audits. There are, however, subtle differences in the approach to internal audits that can have significant effects on implementing and conducting them. (The similarities and differences can be seen by comparing Figure 11.2, the function tree for implementing an internal quality audit, with Figure 11.1, the function tree for implementing an external one.) There are also differences between different internal audits, depending on the modus operandi of the organization being audited and the philosophical approach taken to the quality audit function.

11.3.2   Internal quality audit philosophy

11.3.2.1   Minimal approach.   As indicated in Appendix 1A, most major procurement quality standards either explicitly or implicitly require internal quality audits. But these requirements provide the minimal approach to the technique and barely scratch the surface of the full value that such audits have to offer.

This minimal approach usually requires a review of the full quality system once or twice in a 12-month period. Those standards that specify a period limit the maximal time between reviews to 1 year.

As with external quality audits, an actual internal audit can be carried out as a complete system audit at a particular point in time. This will provide data on the status of the quality system at that point in time, but it may not be representative of its long-term operations. Prior notice must be given that an audit of this nature will be taking place, and individuals will be diverted from their normal tasks to participate in the audit as soon as it starts in their area. Moreover, it might be very difficult for employees to quickly correct a shortcoming despite that prior notice so it will not be noted by the auditor, but the shortcoming's visibility can be obscured. And all trace of randomness or spontaneity in the work area is lost.

The value of this minimal approach can be increased, however, by spreading the system audit over the full amount of time allowed, i.e., 6 or 12 months. The period involved should be divided by the number of functions or elements to be audited to provide the average spacing in time between audits. Randomness of the sampling can then be introduced in two ways:

1. By not announcing the sequence in which the audits are to be conducted, while developing that sequence in such a way that a pattern does not become evident—for example, avoiding starting with order receipt and following the design and production cycle through to shipment

2. By starting the audits at staggered times within the time period allowed

In this way, the status of a particular activity will be known at a point in time, and the results of the audit activities will indicate more fully the status of the overall system in its longer-term operations as the system's various interfaces are examined and evaluated.

11.3.2.2 Internal quality audit and participatory quality assurance. There is a growing recognition that a quality system is not an add-on activity required to "control quality." The modern quality system is accepted as part of an integrated management system for an organization that provides products or services, based on a realization that a quality product results from the following sequence of activities:

1. Knowing the needs of the customer or marketplace

2. Determining how to satisfy those needs, i.e., how to design a suitable product or service

3. Determining the resources, facilities, staff, etc., necessary to provide the end product or service

4. Obtaining the necessary resources, facilities, staff, etc.

5. Providing the desired product or service

6. Determining the problems associated with providing the product or service

7. Determining the acceptability of the product or service from the user's point of view, including desires regarding that product or service which have not been met

8. Improving the design of or methods of providing the product or service to increase its acceptability

9. Repeating steps 6, 7, and 8 on a continuing basis to improve the efficiency and acceptability of the product or service

Notes: (1) The term "needs" denotes the requirements of the customer or marketplace that must be met if the product or service is to be acceptable. (2) The term "desires" denotes the nonmandatory parameters or characteristics of the product or service that would enhance its value to the customer or marketplace.

The first five steps listed above are actually processes that should be subject to the normal, continuous cycle of process control:

1. Knowing the requirements

2. Knowing the capabilities of the process

3. Implementing a process that will meet the requirements

4. Verifying the performance

5. Improving the process to reduce the spread of results

Despite the requirements of some procurement quality standards, the individuals performing an operation or process are in the best position to verify and improve it. These operators are responsible for achieving the desired quality level and yet are seldom provided with the facilities for confirming their performance and making the necessary corrections to the process that would enable them to achieve that quality level. Instead, their supervisors are frequently held responsible and accountable for the quality of the work they do. But if the operators were made responsible for both performing the work and demonstrating that it had been done correctly, this would mean an inspection operation of some kind would be required to verify the performance. And this should be carried out by someone other than the original operator, i.e., by a separate inspector, the individual doing the following operation, etc.

If an integrated quality system designed to satisfy the needs of the customer is introduced, the individual operators and their supervisors are just as involved in the quality process as any member of a formal quality organization, and quality takes its place as first among equals in regard to production concerns. However, a dichotomy arises here: The supervisor of the productive activity, as a member of a commercial organization, must resolve the apparently conflicting requirements of quality, cost, and schedule. It is this dichotomy that has caused concerns by management and standards writers about the degree of objectivity that can be maintained by quality inspection personnel reporting to the supervisors responsible for the work being inspected.

These concerns about objectivity can be answered by the use of one tool of internal quality audits, namely, decision sampling (discussed in Section 10.2 in more detail). Initially, decision sampling can be done on a daily basis in production areas to establish confidence in the ability of the individuals working in those areas to make valid quality decisions. (Note that both accept and reject decisions should be evaluated since both must be valid quality decisions.) Once such confidence has been established, the periods between samplings can be extended. A quality audit is concerned with people and their ability to make valid quality decisions, not with the detailed acceptability of the product or service provided. If the decision-making individuals are making valid quality decisions, the acceptability of the product or service is assured. Thus decision sampling provides positive reinforcement for the line supervisors and their personnel.

In parallel with decision sampling, internal conformity quality audits can evaluate a quality system as it is being applied in practice. This kind of audit determines the conformity between the defined work methods and the effectiveness of those methods. When carried out in relation to decision sampling, a conformity audit can be spread out over a period of 3 months, with its elements being monitored during the decision sampling visits.

By combining decision sampling with conformity quality audits, management can be assured that:

1. The outgoing product or service is meeting its requirements.
2. The methods of providing the product or service are conforming to the defined techniques and that these techniques are efficient.

Thus an internal quality audit provides positive reinforcement to those engaged in the productive activities, along with assurance to management, the customer, and the marketplace that the product or service being provided is satisfactory. In this way, an internal quality audit becomes a natural partner to participatory quality assurance programs.

11.3.3 Initiation of an internal quality audit
program

The executive of the organization, or the management team of the particular organizational element concerned, must establish a policy and general operating rules defining the responsibilities and reporting lines for an internal quality audit program before it is initiated. These may involve the evaluation and support of a participatory quality system or of a quality system incorporating a quality function responsible for all verification processes. Thus this same management team is responsible for defining the nature of the quality system, how it is to be implemented, and how its effectiveness is to be measured.

I have found that an efficient internal quality audit system will soon demonstrate its value, even when started as part of a quality function that includes all the verification activities. If the audit function is combined with training for line personnel in regard to quality awareness and principles, participatory quality assurance systems are a natural outgrowth. Undoubtedly it will be found occasionally that a line supervisor is unable or unwilling to accept the additional responsibility participatory systems entail. However, I have found that to be a rare occurrence. Most supervisors are keen to have their operations run smoothly, and what smoother operation can there be than the "right-first-time" approach? In additional, the confidence by management in the supervisors that participatory systems show is normally enough to overcome any hesitation.

Once the type of quality system has been determined, the management team can charge one of its members with the responsibility for developing and implementing it. This individual should be the person who forms the "quality authority" for the organization, i.e., the person responsible for getting quality problems resolved. Whether this puts the audit and verification activities in the same administrative sphere is immaterial, providing the lead auditor has the independence of action and a position in the organization's hierarchy compatible with the responsibilities involved.

11.3.4 Preparation of quality audit
documents

Once the audit function has been authorized, the audit supervisor and the lead auditor should quickly establish the general ground rules that are to apply and the responsibilities for developing the detailed techniques for implementing them. Working papers, quality audit instructions, draft reporting forms, and corrective action forms should be developed along with audit procedures. The working papers will be strictly conformity audit documents and not related to any particular quality system standard. Thus they will be in the form shown in Appendix 9C or 9D rather than the form shown in Appendix 9A or 9B.

The sampling for determining conformance should be based on completing the system audit in a given period of time, e.g., 3 months. The sample size for particular quality system elements should be spread over this period and over all the departments in which they are applicable. The acceptable performance level (APL or AER) must be chosen to relate to the criticality of each quality system element and the product or service involved. Remember, error-free performance is the only truly acceptable type of performance; anything less is only for determining sample sizes. In this way, the combined sample size will be sufficient to give a high degree of confidence in the results. Similarly, working papers for decision sampling should be prepared, since this technique should form part of the audit program whether results from it are used to authorize movements of goods or not.

Corrective action forms and report forms will be dealt with in Chapters 13 and 14, respectively. Suffice it at this point to indicate that both documents must be couched in terms that are readily understandable by their recipients. In the case of reports to the management team, the team's reporting philosophy and its desires in regard to frequency and general format of reports should be determined and used as the foundations for such reports.

11.3.5 Opening management meeting

This meeting should include all members of the management team, whether line or staff personnel. Its purpose is to reach a *consensus* or *mutual agreement* on how the audit will operate, its lines of responsibility and accountability, the freedom it has to pursue different issues, and constraints, if any, etc.

The presentation by the manager responsible for the quality audit should cover the philosophy and modus operandi of the audit. Once the policies, principles, and philosophy have been agreed on, the audit methods should be reviewed in sufficient detail to permit a full understanding of how the audit activities will be conducted.

It is extremely important that each manager be fully aware of his or her responsibilities under the audit program, as well as under the quality system. Please note that sometimes people do not understand their system responsibilities until their audit program responsibilities are made clear. But an understanding of both, as well as an awareness of the commitment of management to the quality program in all its ramifications, by every echelon of the organization is essential to ensure the program's success.

After the opening meeting, there undoubtedly will be some modifications made to the instructions, procedures, etc., to reflect the decisions made at it.

11.3.6 Operational meeting

Once all of the procedures, instructions, forms, etc., are ready to use, a meeting should be held with line supervisors to review how the audit will operate. More than one meeting may be necessary to keep attendees to a reasonable number. If more than one meeting is held, every endeavor should be made to have a cross section of supervisors at each, so that participants become aware of the interface activities that can result from the audit. All of the auditors should attend the meetings so they can be made aware of any concerns supervisors may feel.

In convening this type of meeting, care must be taken to ensure that no manager's responsibility for downward communication is usurped. The meeting's purpose is to brief managers on how the audit will work. Auditors and management should be prepared to answer queries in such a way as to demonstrate the importance they attach to supporting the activities of line personnel, as well as their awareness of and approach to quality. Auditors and management should also make it clear that no report on any shortcomings discovered during the audit will go to a higher level of management before the supervisors involved have had a chance to review and respond to the report.

It is important that a good working relationship be developed with line supervisors at the start of the audit program and maintained throughout the continuing life of it. The success of an audit depends on a free interchange of ideas and facts between individuals. If the relationship between auditors and line personnel is poor, this interchange will deteriorate and important facts may be obscured or missed.

11.3.7 Initial visits to plant areas

As with external quality auditors, the initial approach by internal auditors to a work area should be through the supervisor concerned. In this case, no introductions to the supervisor are necessary; however, it is a good idea if the supervisor takes the trouble to introduce the auditor and explain the purpose of the audit to those performing the various tasks to be audited. If more than one auditor is involved with a given area, they should all attend this initial visit together so only a single interruption of the line activities occurs. The visit should be scheduled for a time convenient to the supervisor. Both the supervisor and the auditors should be prepared to answer questions about the audit, the quality standards, etc., so that all personnel are made fully aware of the audit's implications. This is particularly important with respect to decision sampling. Owing to the nature of this first visit, it is quite likely that no work elements will be audited during the course of it.

11.3.8   Subsequent visits to plant areas

On subsequent visits to the work area, the auditor should be free to come and go as necessary to perform the audit duties. At times this may involve questioning employees or monitoring the activities of operators or others carrying out their regular duties. Again, the auditor should have full freedom of access to activities, records, etc.

On these visits, the auditor will carry out the various checks required by the working papers, but also will be using all his or her senses in order to determine any irregularities in the activities being monitored. If irregularities are noted, they are subject to recording and analysis, whether they are specifically covered by the working papers or not.

It is important that the auditor keep a record of the various checks and their results as well as of any other observations. These records must be made at the time the audit activity is being carried out, not later in the day from memory. Along with my colleagues, I have experimented with various methods of making these records. We tried clipboards both with blank pages and with preprinted pages, but found the clipboard size awkward for handling on audit rounds. Then we tried pocket-sized monthly diaries where each day had a double page for entries, with times to show when the observations were made. We found these did not get in the way and were easy to use for recording the working notes on audit checks and their results. The data was also easy to transfer to a more permanent report form at the end of each day.

We have also found that having an instant camera, complete with a negative-producing film pack, is a great on-job help for auditors. Frequently, the picture taken is truly "worth a thousand words" in describing a particular problem. If housekeeping is a problem and a subject of the audit, just carrying a camera through the area being monitored can motivate a cleanup.

At all times an auditor must be careful that his or her comments to individual operators do not imply any change in the instructions given them by their supervisor.

11.3.9   Quality auditor reviews

11.3.9.1 Purpose. Internal quality auditors normally work on their own. However, in larger facilities there may be several auditors working in different areas at the same time. This can result in similar audits being conducted in different locations simultaneously. By comparing the information derived from these audits, patterns can sometimes be detected much more quickly than if the results of each location audit are studied separately.

11.3.9.2 Daily reviews. For this reason, a daily review of audit findings should be carried out. This is best done by bringing the auditors together either at the start or the end of each day. Then the auditors can quickly review the current or previous day's findings. An extension of this procedure is the composite record of audits described in Chapter 14. In the latter case, each auditor records the day's results on a summary sheet that covers the whole plant or a major segment of it, e.g., all manufacturing areas.

A daily review of this nature helps ensure that:

1. Each auditor is aware of any problem arising in the facility that could affect the next audit to be conducted.
2. Any problem patterns are detected as soon as possible.
3. Each auditor is able to profit from the experiences of other members of the audit team.

11.3.9.3 Longer-term reviews. Periodic status reports are normally issued by the internal quality auditor(s). These provide information on the audit areas covered during a given period, including the results of the audit activities. The periodicity, along with the format to a large extent, will depend on the size of the organization and the desires of the management team receiving the reports. They may be issued weekly, monthly, quarterly, and/or annually (for more details, see Chapter 14).

Status reports should be prepared by the supervisor of the audit function. However, they should be reviewed with the individual auditors prior to being distributed. In this way, auditors and management alike can be sure of their accuracy and implications.

11.3.10 Reviews with supervisors

The status report on a given line or staff function should be reviewed with the supervisor concerned prior to its being distributed to higher levels of management. In this way, the supervisor will be aware of the report's contents and have the opportunity to prepare for any questions or comments that may result. Such reviews are not held to reach a consensus on the wording of status reports, but rather to ensure that the supervisor knows what is being said in them. If some pertinent comment comes out during the review, it is up to the auditor(s) to decide whether reference should be made to it in the report.

11.3.11 Managerial reviews

In addition to status reports, reviews of quality audit activities and their results should be held periodically with the management team. The periodicity involved will depend on the desires of those concerned. However, it should not be longer than 1 year.

These reviews will enable the management team to judge whether or not the audit activity is achieving the desired results. At the review meetings, it may be decided to extend the responsibilities of the audit, to increase or decrease its attention to various areas, or to make other changes in order to improve its value.

An assessment of the overall effectiveness of the quality system should form part of these reviews. This step is included not because it is required by many procurement quality standards, but because it is good management practice to review all programs or systems periodically. It also helps demonstrate the continued commitment of management to the quality system.

11.4 Implementing Quality Audits in a Service Industry

11.4.1 Introduction

A quality audit is the best technique for determining the effectiveness of a quality system in any service industry organization. In some instances, it is the only way to monitor the quality of the service itself, as well as the decision-making activities that make up the quality system. However, there are virtually no national or international quality system standards for service industries.

In general, these audits will be conducted to answer one or more of the following questions:

1. Does the quality system for the organization, e.g., a hospital or other medical institution, meet the requirements of the applicable government regulations, statutes, etc.? If it does, this audit may lead to the registration or certification of the facility by an approval or licensing agency.

2. Are the activities, functions, operations, etc., effective from an operational point of view and do they follow the defined methods, techniques, etc.?

3. Do the results of the organization's activities fulfill the expectations of customers and the marketplace?

11.4.2 Suitability quality audit

This type of quality audit is normally conducted to provide the answer to question 1 above. This means a suitability quality audit. And the initiation and implementation of a suitability quality audit in a service industry organization follows basically the same steps as one carried out in a manufacturing organization. These audits use the working papers described in Section 9.4.6.7, and the audit activity is implemented using tactics similar to those outlined in Section 11.2.

11.4.3 Conformity quality audit

A conformity quality audit of a service industry organization in regard to its operational effectiveness and conformity to stated procedures and other documented requirements follows the same steps as outlined earlier for such an audit of a manufacturing organization. The working papers are prepared as discussed in Section 9.4.6.7, with their details depending on the applicable procedures, instructions, and other documentation. Their format is similar to that shown in Appendix 9C or 9D. In addition to working papers, decision sampling (see Section 10.2) is a major control technique.

Whether external or internal auditors are used depends on the size and complexity of the establishment and its relationship to other organizations. Many large establishments such as hospitals, major hotels, banks, airlines, etc., probably have resident auditors to conduct quality audits on a continuing basis. Smaller establishments such as fast food outlets, small hotels, etc., belonging to a chain or franchised from a holding company, are usually audited by personnel working out of the headquarters of the parent organization or out of a district office. In either case, the auditor evaluates the various activities and quality decisions to determine the effectiveness of the quality system and its conformity to the documented requirements. In these cases, working papers are based on documentation provided by the organization's headquarters that the outlet is supposed to use.

11.4.4 External quality audits of a service industry from the customer's perspective

A quality audit of this nature would be conducted by individuals not known in the facility but using the service as a normal customer would. The auditor may be part of a special group operating out of a headquarters or district office or part of a third-party auditing organization conducting the evaluations under contract to the service company. Working papers are prepared in the same manner as already discussed in Section 9.4.6.7, and a typical set of instructions is given in Appendix 9E. The working papers must cover the expectations of the customer as seen through the eyes of the market research department and its various surveys.

Implementing this type of audit does not require the help of anyone in the facility to be evaluated. The auditor will visit the facility and use its services as an ordinary customer. While using the services, the auditor should use all of his or her senses, particularly peripheral vision, to monitor what is happening in the various areas of the facility. Although instructions will outline guidelines for the auditor, these should not be considered as rigid rules, since unplanned occurrences can provide valuable information for this type of audit.

Normally the data derived from this type of visit is combined with the results of questionnaires provided to actual customers. The questionnaires ask the customers to rate the various services in terms of how satisfactorily they are provided.

11.5 Conclusions

All quality audits should be constructive activities that will be welcomed by those being audited as positive reinforcements of their own activities. They should allow the auditee's staff to demonstrate pride in their work and in the organization's quality system. The proper implementation of both external and internal quality audits greatly assists in this reinforcement.

An external quality audit also serves as a training ground for the auditee organization's guides, since it gives them the opportunity to see their company's quality system through the eyes of an independent observer. This opportunity will frequently lead to improvements in the system, even where no adverse observation is made. For example, problems in demonstrating a particular work element may reveal a shortcoming in the system no one was aware of previously.

Chapter

12

Analysis of
the Quality Audit

12.1 Introduction

Analysis is the conversion of raw data into useful information. This information can then help in managing and improving the performance of the quality system being audited, whether the information has to do with the overall organization, a particular product, or a specific service or process.

Quality audits are conducted to determine one or more of the following:

1. The suitability of an organization's quality system, i.e., the conformance of the quality program to the requirements of the reference quality standard

2. The conformity of the acts or decisions of an organization's personnel to the requirements of the quality program as defined in the quality policies, quality manual, quality procedures or instructions, quality standards, product documentation—i.e., drawings, specifications, standards, regulations, etc.—process procedures and instructions, and so forth

3. The effectiveness of the various activities making up an organization's quality system

The raw findings or observations noted during the audit will show, in their broader interpretation, that:

1. The quality program does or does not conform to the reference standard.

2. The acts or decisions by personnel do or do not conform to the quality program.

In cases where no nonconformities or departures from the reference documents are noted, these generalities may be acceptable. Occasionally they may even suffice when nonconformities are detected. But in neither case would they reveal the actual effectiveness of the quality system. Analysis converts the raw data of findings into collated information on what the actual quality performance is and what actions should be taken to improve it.

Moreover, normally both client and auditee would like to know how effectively the system is performing and how that performance rates against objectives or expectations. Since performance is normally measured quantitatively, the raw data must be converted through analysis into this form.

Any improvement in an operating system requires:

1. The present level of performance to serve as a benchmark for later measurements

2. Information on the cause of any shortcomings in the existing system in order to eliminate them

Thus the level of performance and the causes of problems must be determined through analysis of data on symptoms or effects noted during the audit. From this information, improved methods, techniques, etc., can be developed to overcome existing shortcomings.

In addition the analysis must provide some means of confirming that the changes introduced will actually result in a system improvement. It must also provide on-going information about system performance so that any trends developing in the system can be determined. The quality system itself, similarly to the quality of a product or process, must be subject to continuous improvement if it is to achieve its objectives. Without continuous improvement, the quality system will become stagnant and gradually deteriorate in effectiveness.

Thus the analysis of quality audit findings falls into several categories, illustrated in Figure 12.1, including:

1. Qualitative analysis of patterns detected in printouts, summary reports, etc.

2. Problem solving to separate cause from effect or symptoms in order to determine what needs to be corrected to prevent a problem from recurring

3. Quantitative analysis of findings to provide performance figures for the various activities within the quality system

4. Quantitative or qualitative information on the trends of performance in certain key areas

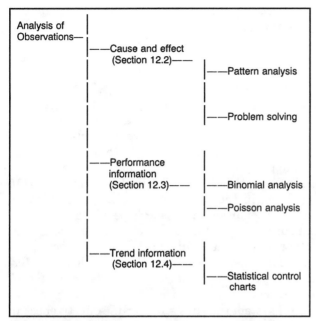

Figure 12.1   Function tree for the analysis of quality audit findings.

12.2   Cause and Effect Analysis

12.2.1   Introduction

Cause and effect analysis is used to separate the element of the quality system needing correction from the symptoms or visible results of the shortcoming. A quality audit points out the symptoms or visible results of any shortcoming. Analysis of the data identifies the cause of the shortcoming.

12.2.2   Pattern analysis

Pattern analysis involves the collation of data in a manner that readily reveals any kind of clustering that occurs. This technique is of major value in internal quality audits, since it is so effective in making use of data from repetitive audits. Thus it can be location- and time-sensitive. It is of limited value in external quality audits owing to the lack of repetition in such audits.

12.2.2.1   Matrix presentation.   One method of pattern analysis is to plot a matrix of quality elements in different locations in the facility against time. The plot should show the status or result of the audit under four gradations: satisfactory, inadequate, adequate, and unsatisfactory. It may also be necessary to show locations not being audited. Clustering or patterns may occur horizontally, vertically, or overall. A typical format for this type of report is shown in Section 14.3.2. Suffice it at this point to discuss Table 12.1, an extract of a summation sheet dealing with document control, where the results are shown as satisfactory (OK) or not satisfactory (NOK).

Examination of Table 12.1 shows two basic patterns, one location-sensitive and the other time-sensitive. Without examining the problem in detail, it cannot be said whether these are interrelated or not. The horizontal NOKs in the "Shop C" now indicate a problem with document control that appears to be peculiar to that shop, as all other areas generally performed satisfactorily. Similarly, during week 10 something drastic occurred in the document control system that resulted in an unsatisfactory condition in all the shops. If the problem in Shop C during that time period is the same as it was in the earlier periods, a possible cause might be the failure of "corrective action" to provide an improvement in the controls there. If the problem is different, though, the overall problem in week 10 results from a self-contained cause. A scattering of NOKs over the whole matrix indicates a document control system that has a borderline effectiveness in its operation. Particular weaknesses will reveal similarities of cause.

12.2.2.2 Computer-generated patterns. With an internal quality audit, each observation of a shortcoming will give rise to some form of corrective action request. These requests are discussed in more detail in Chapter 13. But I want to note at this point that the key pieces of information on the request are:

1. The cause of the problem

2. The supervisor, or the location, responsible for taking the necessary corrective action

3. The drawing or part number for the items involved in the problem

TABLE 12.1   Extract from a Quality Audit Composite Summary Sheet—Document Control

Element and location	Week numbers					
	7	8	9	10	11	12
Shop A	OK	OK	OK	NOK	OK	OK
Shop B	OK	OK	OK	NOK	CK	OK
Shop C	NOK	NOK	NOK	NOK	OK	OK
Shop D	OK	OK	OK	NOK	OK	OK
Shop E	OK	OK	OK	NOK	OK	OK

4. The date or equipment serial number by which the correction must have been made and verified

Where suitable computer facilities are available, this data should be entered into a data bank, along with any additional data that may be required for more detailed analysis.

The simplest forms of computer-generated patterns will result from alphanumeric sorting and counting. This process normally takes two forms:

1. An examination of the problem or elements of the quality system that caused the observation, along with a supplementary examination of the locations involved with the major causes

2. An examination of the locations responsible for the corrective action, with a supplementary examination of the quality system elements involved

A computer program would use the following logical sequence of steps for the first of the above instances:

1. Arrange the prime sort into alphanumeric sequence. In the first instance, this would amount to examining each cause in turn and rearranging the lines of data until the causes were in alphanumeric sequence. (In the second instance, the prime sort would be location- or supervisor-oriented.)

2. Count the number of like-cause entries and record them by name and quantity.

3. Arrange the data from step 2 into descending numbers of occurrences, i.e., according to Pareto distribution.

4. Take the highest numbers of causes either as a percentage of occurrences or as an absolute value, and perform a second alphanumeric sort on the secondary data, i.e., the location, for each cause.

5. Within each cause, count the number of like-location entries and record them by name and quantity.

6. Within each cause, arrange the locations responsible in descending numbers of occurrences.

7. Plot a Pareto distribution of the causes selected in step 4.

8. Print the data derived from step 7, down to a preselected minimum number of occurrences.

9. Repeat the sequence with locations as the prime sort and causes or quality elements as the secondary sort within the major locations.

This fairly simple data processing method will provide graphic patterns enabling the analyst to concentrate on the areas of major concern. It will also provide a listing to assist in the diagnostic analysis. Similar programs can be used to provide:

1. Similar data with respect to products being audited
2. A listing of corrective actions that are needed and those outstanding

This kind of use of a computer eliminates the drudgery from analyses by storing and collating data into usable form, enabling the analyst to concentrate on the analysis itself.

12.2.2.3 Bar graph presentation.

Early on in an audit, two kinds of bar graphs should be prepared for the presentation of audit data. The horizontal axis of one should be labeled in terms of the operating functions or locations being audited and the horizontal axis of the other labeled in terms of the elements of the quality system being monitored. The vertical axes on both should be labeled numerically, for recording the numbers of observed shortcomings noted at each location or for each element. See Figure 12.2.

When an observation has been made and a corrective action notice issued, a notation should be added to each chart, indicating the responsible functional or operating location in one case and the problematic system element in the other. In both cases, the number of the corrective action request should be entered, to simplify subsequent analysis. As additional observations are made, they should be entered in the same way. If an element of the quality system develops sufficient observations, it may be necessary to start a new graph for that element alone, with the horizontal axis labeled in terms of the particular causes that have been identified.

Pareto-type analysis can be applied to these bar graphs without redrawing them by determining the location or element with the most occurrences. These can then be analyzed in more detail.

A fairly simple binomial calculation can be used to determine the probability of having the number of occurrences noted on a given bar. The average error rate, the p of the binomial expression, can be calculated for the given bar from the data available. The number of audits concerned with the bar times the sample size per audit gives the total number of checks made, the n of the binomial expression. The number of observations plotted gives the value of c. From these figures, the probability P_c of having the number of incidents shown on the bar can be calculated using the following equation:

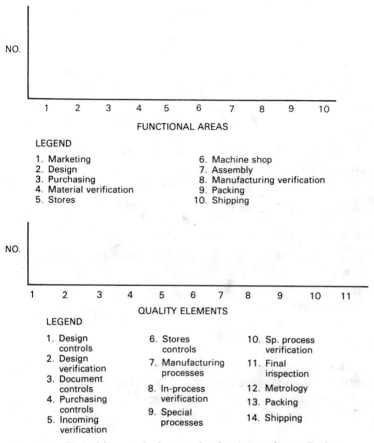

LEGEND

1. Marketing
2. Design
3. Purchasing
4. Material verification
5. Stores

6. Machine shop
7. Assembly
8. Manufacturing verification
9. Packing
10. Shipping

LEGEND

1. Design controls
2. Design verification
3. Document controls
4. Purchasing controls
5. Incoming verification

6. Stores controls
7. Manufacturing processes
8. In-process verification
9. Special processes

10. Sp. process verification
11. Final inspection
12. Metrology
13. Packing
14. Shipping

Figure 12.2 Typical layouts for bar graphs showing quality audit data.

$$P_c = \frac{n!}{c!(n-c)} p^c q^{n-c} \tag{12.1}$$

where P_c = probability of exactly c occurrences

n = number of samples taken in the audit

c = number of observations made of shortcomings

p = average error rate for the population

= c/An where A denotes the number of audits

q = average performance level for the population

= $1 - p$

If the event is rare, e.g., < 5 percent, some unusual cause is involved. If, on the other hand, it is not a rare event, the number of observations is within the realm of expectations for the error rate calculated.

12.2.3 Problem solving

Many texts and seminars dealing with problem solving are available. These are effective to varying degrees, but all of them address the key issue of identifying the cause of a problem based on the symptoms or observations detected.

One of the most effective techniques is the Ishikawa cause and effect diagram discussed in Section 8.8.

12.3 Performance Information

12.3.1 Introduction

The derivation of quantified data is essential if performance measurement is to be meaningful. Without a numerical benchmark or reference, it is impossible to determine if "improved processes or systems" actually improve performance.

12.3.2 Binomial analysis

The binomial expansion permits the performance level to be accurately quantified when shortcomings have been observed. The known factors are the sample size n and the number of occurrences or observations c. To a first approximation, the performance level can be found as follows:

$$\text{Average error rate} = \frac{c}{n} = p$$

or

$$\text{Average performance level} = 1 - p = q$$

These two values have a 50 percent probability of occurring, i.e., the confidence level in the figure derived is 50 percent. This is not normally an acceptable confidence level for measuring performance levels.

Since a quality audit operates in a binary mode, it measures performance through the number of errors detected within a given sample size. The binomial expansion permits the calculation of the probability P_c of finding exactly c observations in a given sample size n when the nonconformity rate p is the average error rate. This calculation uses Equation (12.1).

The average error rate p can also be calculated for known values of n and c and particular values of the confidence level P_c using Equation

TABLE 12.2   Table of Performance Levels for Different Sample Sizes, Number of Observations and Confidence Levels

Sample size	Number of observations			
	Zero	One	Two	Three
	90 % Confidence Level			
50	95.499	92.998	90.767	88.661
100	97.724	96.461	95.337	94.280
200	98.855	98.220	97.657	97.127
	95 % Confidence Level			
50	94.184	91.279	88.719	86.319
100	97.049	95.569	94.265	93.041
200	98.513	97.766	97.108	96.491
	99 % Confidence Level			
50	91.201	87.751	84.751	81.991
100	95.499	93.899	92.154	90.674
200	97.724	96.798	96.002	95.265

(12.1). However, determining p for a given value of P_c is more complex. Nevertheless, an iterative computer program can perform the required calculations to the desired degree of precision and provide tables from which the following type of statement can be made:

With a confidence level of P_c, the quality audit shows that the actual performance level is no worse than x percent.

x is the value of p determined by the above calculation.

The iterative program uses the following logic:

1. For the desired confidence level P and with zero nonconforming observations, calculate the corresponding error rate p for the given sample size n, using the expression $P_0 = p^n$ or by transposing $p = (P_0)^{1/n}$.

2. Using the above values p and n, calculate P_c using Equation (12.1).

3. Repeat number 2 incrementally increasing or decreasing the value of p until the desired value of P_c is obtained. The final value of p is a performance level corresponding to those shown in Table 12.2. *Note*: In practice it is necessary to place a tolerance on the acceptable value of P_c to limit computer cycling time.

This calculation can be performed on an audit-by-audit basis and used to plot a statistical control chart (see Section 12.4.3). However, where an internal quality audit is being repeated on a regular basis, the cumulative values of the sample size n and the number of observations c can be used to provide a more accurate average performance level. A sampling of the values obtained by iteration is shown in Table 12.2. They have been selected from the overall tables for confidence levels of 90, 95, and 99 percent.

The confidence level is the key piece of information determining the performance levels of the quality system. As the confidence level increases for given numbers of sample and observations, the higher limit of the error rate gets higher, or conversely the lower limit of the performance level gets lower.

Example   For an audit sample of 50, two observations were made of shortcomings. Based on this data the following statements could be made:

1. With a 50 percent confidence level, the average error rate in the area concerned is no higher than 4 percent.
2. With a 90 percent confidence level, the average error rate in the area concerned is no higher than 9.233 percent.
3. With a 95 percent confidence level, the average error rate in the area concerned is no higher than 11.281 percent.
4. With a 99 percent confidence level, the average error rate in the area concerned is no higher than 15.249 percent.

12.3.3 Poisson analysis

A similar approach can be developed for the Poisson distribution, to determine the performance levels that result in selected values of the probability of occurrences for given sample sizes and numbers of observations. This is an outgrowth of the discussion in Section 10.5.

The formula for the Poisson distribution is:

$$P_c = \frac{(np)^c}{c!} e^{-np} \tag{12.2}$$

where P_c = probability of c occurrences
n = sample size
p = average error rate or nonconformity rate
c = number of occurrences
e = exponential base = 2.718281

Substituting $w = np$,

$$P_c = \frac{w^c}{c!} e^{-w}$$

And transposing, we have

$$c!P_c = w^c e^{-w} \tag{12.3}$$

The solution of Equation (12.2) or (12.3), for other than $c = 0$, requires an iterative computer program that will refine the values of np for a given number of occurrences until the desired precision of the

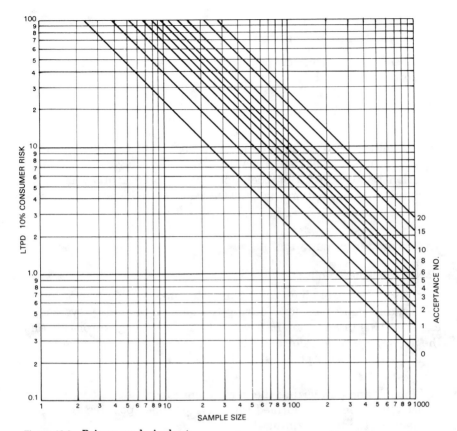

Figure 12.3 Poisson analysis chart.

confidence level has been obtained. To do so the iterative program varies the value of p in ever smaller increments until the desired value of
P_c is obtained.

From these calculations of np for different confidence levels and
numbers of occurrences, a graph can be plotted for each confidence
level similar to the one shown in Figure 12.3 for a 90 percent confidence level. This chart was originally developed for an LTPD sampling plan. In applying it to this problem, the following steps are
taken:

1. Find the sample size on the horizontal axis.

2. Move up from that point to the intercept of the vertical and the diagonal for the acceptance number c corresponding to the number of
   occurrences noted.

3. Move horizontally and read the value of the performance
   levelfromthe vertical scale.

4. Check the value on the vertical scale, which gives you the error
   rate you can use in a statement such as the following:

   > There is only a 10 percent probability that the error rate in the area con-
   > cerned is higher than x percent.

12.3.4 Conclusions

Either the binomial or the Poisson equation may be used for determining performance levels or error rates. However, I favor the binomial
equation because of its broader application to data analysis where
there are only two options, i.e., OK/NOK, GO/NOGO, etc.

12.4   Trend Information

12.4.1   Statistical control charts

A statistical control chart can be used to make visible trends in the
performance data measured during the audit of a quality system. It
can show this information in terms of both the average error rate and
the acceptable performance level. But when deciding on which factor
to plot, the following points should be borne in mind:

1. Plotting and publicizing the performance level shows how well an
   activity is being performed and thus provides positive reinforcement for those performing it.

2. Plotting and publicizing the error rate highlights the mistakes that
   have been made and thus provides negative reinforcement unless
   the facility being audited is actively pursuing a zero-defect program.

Virtually any of the traditional types of statistical control charts for
attributes can be used to plot quality audit data. But some of them
present problems. For instance, the performance level and the error
rate are attributes measurements and can be plotted in terms of p
charts: the error rate as the normal p of such a chart, and the performance level on a variation of the p chart, where $1 - p$ is plotted instead of the usual p.

However, on p charts, the graphed line slopes downward as the system is improved and the error rate is decreased. The error rate should
be low and remain low unless some serious shortcoming occurs. Thus
normally the graphed line quickly becomes asymptotic to zero, unless
a very expanded scale or a logarithmic one is used for the vertical

axis. But neither of these lend themselves to easy visualization of the performance.

Similarly, the graphed line showing performance level, i.e., $1 - p$, under regular conditions quickly becomes asymptotic to unity, and thus creates the same problems in relation to scale and visualization.

Moreover, psychologically speaking, most people expect a graph to show improvement by an upward slope and degradation by a downward one, no matter whether performance or error rates are being plotted. This leads to the question: Do I plot the error rate directly and explain that a downward slope is an improvement, or its reciprocal to achieve a psychologically acceptable upward slope?

After experimenting with various graphical representations, I have found CuSum control charts to be the most acceptable, both for auditees and for auditors, in plotting audit data. On them is plotted the cumulative sum of the differences between successive audit readings and a preselected normal value, frequently referred to as the "norm." Traditionally that norm is the $\overline{X}$ or the nominal value for a process.

In developing a CuSum chart for an audit application, the following sequence of events should be followed:

1. Calculate the individual performance level at the desired confidence level for each of the first 10 audits of the area or activity concerned.

2. Calculate the average performance level for the 10 audits. This becomes the norm for the chart.

3. Calculate the algebraic difference between the average performance level and the individual performance level for each audit. To maintain an upward slope for improved performance, the average performance level is subtracted from the individual performance level.

4. Begin plotting. The first point is the difference value for the first audit.

5. Continue plotting. The next point is the algebraic sum of the first two differences.

6. Plot successive points by adding the new differences to the previous sum. (Hence the name "CuSum.")

7. As the plot nears the end of the time scale, calculate a new norm for the next sheet by taking the slope of the graph and either adding that increment to the norm for an upward slope or subtracting it for a downward slope.

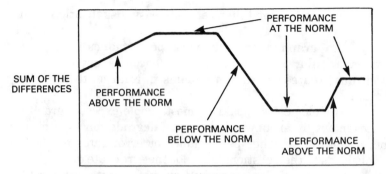

TIME SCALE

Figure 12.4   Interpretation of CuSum statistical control charts.

Note: Plotting the trend of error rates will follow the same principles just outlined, except for step 2. In step 2, the individual error rate will be subtracted from the norm or average error rate. This will give an upward slope for reductions in error rates, i.e., for improvements in the system operation.

Thus CuSum charts avoid the problems associated with p charts since interpreting them in an audit application is relatively straightforward (also see Figure 12.4):

1. An upward slope indicates an improvement in the process.

2. A downward slope indicates a degradation in the process.

3. A horizontal line, regardless of its position on the chart, indicates a steady performance at the norm.

However, please note that, although the actual performance is the same, depending on its relationship to the previous plots, a horizontal line could indicate either an improvement or a degradation from earlier performance. If it follows a downward sloping line, it shows an earlier degradation has ceased and the process is now performing at the norm. If it follows an upward sloping line, it indicates a degradation from the improvement shown in the previous performance.

Example

The performance levels for 10 individual audits are given in column 1 of the table below. Calculate the points to be plotted on a CuSum control chart for the values given.

Audit number	Individual performance level, X	Individual difference, $\Delta X = X - \overline{X}$	CuSum differences, $\Sigma \Delta X$
1.	95.0	0.0	0.0
2.	94.5	-0.5	-0.5
3.	95.5	0.5	0.0
4.	96.0	1.0	1.0
5.	95.5	0.5	1.5
6.	94.5	-0.5	1.0
7.	94.0	-1.0	0.0
8.	95.5	0.5	0.5
9.	95.0	0.0	0.5
10.	94.5	-0.5	0.0

$$\Sigma X = 950$$
$$\overline{X} = 950/10 = 95$$

12.5 Conclusions

The trends revealed by a quality audit can be shown by a variety of techniques, some of which are discussed above. The method used for reporting should be the one that will be most easily understood by the report's recipients.

I have found that methods showing patterns, especially a summary report sheet as discussed in Sections 12.2.2 and 14.3.2, are particularly effective in dealing with first-line manufacturing supervisors. In one instance, a manufacturing manager used a copy of this report as part of the weekly production review.

For those activities to which process control charts are applicable, the CuSum approach makes performance clearly visible. Similarly in the management and professional areas, CuSum is one of the most effective means of reporting trends. The handling of these latter in reports is discussed in Section 14.3.3.

13

Corrective Action
and Followup with
the Auditee

13.1 Introduction

The prime objectives of a quality audit are to:

1. Make clear to the client the status of a quality program with respect to its meeting the requirements of a predetermined standard

2. Initiate corrective action on any shortcomings detected

In practice, making clear the status of a program includes providing data on any corrective action required and initiated.

Although the auditor is frequently assigned the responsibility for initiating corrective action, it should be recognized that in virtually all cases this role will be: (1) motivating those responsible for the activity to make the necessary changes to correct the observation and prevent its recurrence and (2) Determining the adequacy of the proposed and actual changes. Normally an auditor should *not* suggest the method of corrective action since doing so would involve her or him in the development of system changes. This involvement would in turn disqualify the auditor from assessing the validity of the action, as independence from the process would have been lost.

The activities associated with corrective action are shown in Figure 13.1, in the function tree for corrective action. The principles involved are basically the same for both external and internal quality audits. However, there will be some variations in the recommended techniques.

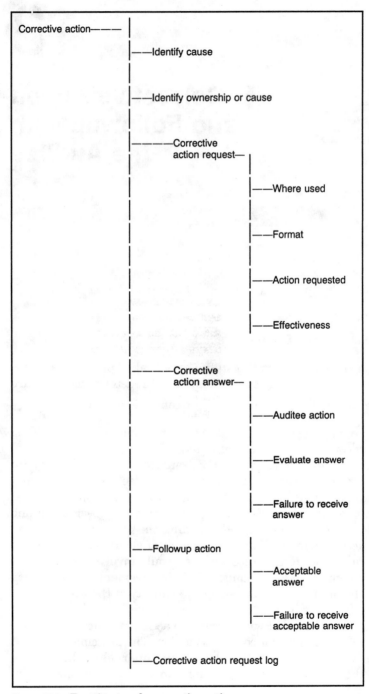

Figure 13.1   Function tree for corrective action.

13.2   Identifying the "Cause" from the Observations

As discussed in Chapter 12, it is essential that the analysis of any observation revealing a shortcoming be thorough in order to segregate the actual problem or cause of the shortcoming from the observations or symptoms indicating that a problem is present.

This analysis may reveal more than one problem related to a single set of data. For example, an audit of product leaving a particular work area reveals that a number of faulty units have been accepted by the outgoing inspection function. This situation raises the following questions:

1. What happened in the production process to cause the faulty product?
2. Why was the faulty product not detected by the operator(s) responsible for carrying out the operation or subsequent operations?
3. Why was the faulty product not detected by the verification system?

Thus a single observation has indicated possible falldowns in the productive as well as the verification processes. Both aspects must be examined each time the need for corrective action is being assessed. Corrective action may be required for both functions, as it frequently is in such situations. Depending on the organization being audited, the activities may be under a single supervisor or under more than one. A number of national and international quality system standards require separate supervision for these functions.

Any corrective action request should also clearly state the requirement which is not being met and the specific nature of the shortcoming.

13.3   Ownership of the Problem

Having determined the exact nature of the problem (its cause), the auditor must determine who is responsible for the activities related to that problem. This "ownership" of the activity may require very careful analysis of the data available. For example, if an operator is having a problem with a process:

1. Does the fault lie with the operator through lack of training or facilities?
2. Are the defined methods or processes inadequate?
3. Is the product design faulty?

The ownership of all these problems might lie with the production supervisor; however, this is rarely the case. For the situation cited, the problems could involve the training department, the industrial engineering department, or the product design department. Similarly, shortcomings in the verification system can involve more than one function.

Requests for corrective action are normally sent to the supervisor responsible for implementing it.

13.4   Corrective Action Requests

13.4.1   Where corrective action requests are used

The client is responsible for defining the extent to which corrective action requests are to be used. The client might only require an audit report indicating the extent of an activity's compliance to the predetermined requirements. The report may be in either qualitative or quantitative terms, but when possible its information should be quantified (see Chapters 12 and 14).

The initiation of corrective action requests, and the degree of detail they contain, will to a large extent depend on the relationship between the client, the auditor, and the auditee. The client's desires will be reflected in the terms of reference given to the auditors. These may be short term, i.e., peculiar to a particular audit, or long term, i.e., covering a continuing operation.

For external quality audits, the content can vary from a simple statement of system nonconformance to a detailed request for specific corrective actions. Statements on nonconformity in quality audit reports are discussed in Chapter 14. Requests for specific corrective actions are covered in this chapter.

For internal quality audits, each shortcoming observed will result in some form of action request. Thus internal audits will always use the techniques discussed in this chapter.

13.4.2   Corrective action request format

13.4.2.1   Corrective action request letter. The corrective action request letter shown in Figure 13.2 is normally used when reporting the results of a quality audit carried out in relation to a potential supplier or vendor. These audits are used to determine the initial suitability of suppliers. A letter of this nature would be used where the auditee had a major program shortcoming and might or might not be interested in taking the necessary corrective action. Once a supplier has been initially approved, shortcomings revealed by future audits would use either quality audit letters or corrective action requests.

P. O. Box abc
Mytown

November X, 19XX

Mr. John Brown, President
A. N. Other Forging Co., Ltd.
P. O. Box ano
Urtown

Dear Mr. Brown:

Quality Audit of Your Facility

Thank you for the courtesy extended by you and your staff during our recent audit of your facility.

As indicated to you at the debriefing meeting, we found your facility did not meet the requirements of our Quality Procurement Standard ABC-JIT-1986 in the following areas:
  1) The control of procurement and the resulting materiel as required by paragraph 3.3 of the above standard.
  2) The implementation of a Quality Audit system as required by paragraph 3.15 of the above standard.
  3) The integration of the Quality Audit into your Management Review of the system as required by paragraph 3.2 of the above standard.

Acceptable corrective action is required in each of the above areas if you are to be placed on our Qualified Suppliers List (QSL).

If you desire to be placed on our QSL, please provide this office with details of the corrective actions taken and when they will be fully implemented. This will enable a followup Quality Audit to be scheduled, in order to evaluate these actions.

Yours sincerely,

I. M. Wright, Manager
Supplier Audit Department
ABC Manufacturing Co., Ltd.

cc: Client
    Designated Quality Representative

Figure 13.2 Corrective action request letter.

A corrective action request letter must clearly state the acceptability or nonacceptability of the auditee's quality system. If it is unacceptable, the areas of shortcoming should also be clearly identified.

Normally a letter of this type is sent to the head of the auditee's organization, i.e., the chief executive officer, division manager, etc. A copy may be included for the organization's designated quality representative.

External auditors should provide the client with copies of all corrective action correspondence with the auditee, whether that client is a member of the same organization or not.

13.4.2.2 Specific corrective action request. A typical corrective action request for use in addressing a specific problem is shown in Figure 13.3. This kind of form is used to request the initiation of action on a particular problem by the individual responsible and accountable for the activity needing correction. Where word processing equipment is

---

A.B.C. Manufacturing Co. Ltd. **Corrective Action Request**	Area 1
To _____ Location _____ Ref. No. _____ Date _____ Product/Process _____ Problem _____	Area 2
Quality Audit has made the following observation in your area of concern. Action is requested to correct these observations and prevent recurrence of them. Requirement _____ _____ Observation _____ _____ Action is required by _____	Area 3
The following actions have been taken to correct this occurence and to prevent recurrence. _____ _____ _____ _____ _____ _____ Signed _____ Date _____ Position _____	Area 4
The actions taken are Acceptable / not Acceptable. Signed _____ Date _____ (Quality Audit) Further Action _____ _____	Area 5

Figure 13.3 Corrective action request form.

available, it is recommended that the request form be designed on and printed out from it. Preferably the equipment should be in the audit office to reduce both delays in preparing forms and the number of possible transcription errors. The forms should always be printed on rolled, letter-quality bond paper.

As shown by the marginal notes in Figure 13.3, this type of request is divided into five areas or action segments. Each area fulfills certain needs in the communication and analysis chain. The exact layout of each area depends on the needs of the particular organization.

Area 1 identifies the organization and the form by name. The name of the organization may be printed out as part of the word processing activity or it may already appear on the paper to be used for the printout. "Corrective Action Request" is the most common title for the form. However, for internal quality audits in organizations where teamwork is stressed and every endeavor is made to avoid interdepartmental irritations, a more suitable title is "Quality Audit Action Request," as shown in Figure 13.4. This wording will reduce some of the confrontational aspects of the request and hopefully eliminate some of the defensive attitudes that can arise when an addressee is requested to take corrective action with respect to some personal action. The remainder of the form will go with either title.

13.4.2.3 Use of forms in problem solving. Many problem-solving techniques use specialized forms to identify and analyze the problem concerned. These can frequently be adapted to serve as quality audit action request forms. Where a large portion of the personnel have been trained in a particular problem-solving technique, this can help motivate a complete analysis. A successful action request form was adapted from the Kepner-Tregoe problem analysis forms. Similar adaptations can be made from your favorite analysis form. However, care must be taken not to infringe copyrights when making such adaptations.

13.4.3 Requested action

Area 2 of Figure 13.3 summarizes data about the request under a series of headings suitable for computer storage and analysis. It may be necessary to convert the lines on which the data is entered into a series of squares to force those making the entries to maintain a consistent format and keep the entries within some specified number of characters.

A.B.C. Manufacturing Co. Ltd.
Quality Audit Action Request

Figure 13.4 Corrective action request form, alternative area 1.

Quality Audit has made the following observations in your area of concern.
The following action is requested to correct these observations and prevent recurrence of them._____
  Requirement_____
  Observations_____
  Action is required by_____

Figure 13.5   Corrective action request form, alternative area 3.

The addressee (the "To" line) should be the individual responsible and accountable for the activity concerned. For internal quality audits this is normally a supervisor. For initial external quality audits, it is normally the company head, i.e., the CEO, division manager, etc. For external quality audits on approved suppliers, it is normally the designated quality representative of the supplier, or in the absence of that position, the company head of the supplier.

In some instances, some form of numeric code may be used to identify the product or process and the problem. However, I favor the use of plain language for each of these headings since computerized alphanumeric sorts are now commonplace. But note: The use of plain language will entail some training to ensure that all auditors use the same terminology.

The "Problem" line corresponds to the heading of a letter—it explains the reason for the request.

Area 3 of Figure 13.3 is the body of the request, describing the requirements that form the reference for the audit, the observations made indicating the nature of the problem, and the action requested. Figure 13.5 shows an alternative layout covering all three points. But no matter which layout is used, it should be noted that the requested action is twofold, involving:

1. The action necessary to correct the immediate problem, i.e., the short-term action

2. The action necessary to prevent a recurrence of the same problem, i.e., the long-term action

The prevention of a recurrence is frequently overlooked amid the pressures of solving the immediate problem. The auditor provides a stabilizing influence in this type of situation.

In summary, area 3 provides the data from which the recipient can determine the necessary actions.

13.4.4   Effectivity

One of the important sectors of area 3 is the effectivity line ("Action is required by"), which tells the recipient when the requested action is to be completed by. The effectivity may be given as a date, a serial number, a batch number, etc. Whichever is used, it must be practical and selected with care. Allowance must be made for factors that affect this requirement, such as the seriousness of the problem, the difficulty of the action requested, possible implementation delays, etc.

13.5 Auditee Action

Area 4 of Figure 13.3 is completed by the addressee. It constitutes the record of the actions taken as a result of the request. It is essential that the answer describe both the short-term actions taken to correct the immediate problem and the long-term actions instituted to prevent a recurrence.

In the case of a general request for corrective action (Figure 13.2), the results of the request will normally be received in letter form, possibly with new procedures, etc., attached. This letter should also cover the actions taken to correct the immediate situation and prevent a recurrence.

But whether the reply is by filled-out corrective action request form or separate letter, it is the responsibility of the auditee to confirm the problem from the symptoms reported, develop short- and long-term corrective actions taken, and verify that the corrective actions have eliminated the existing problem and will prevent it from recurring. Verification data should be provided to the auditor.

13.6 Auditor Evaluation of the Actions Taken

The auditor records the results of the evaluation of the corrective actions taken in area 5 of Figure 13.3. This evaluation may be based on one or more of the following: a followup audit, an analysis of the verification data provided for the actions shown in area 4, the results of special verification tests, etc. As can be seen on Figure 13.3, the auditor's decision takes the form of "acceptable" or "not acceptable."

When corrective actions have been requested by letter, the results of those actions must also be evaluated and a similar decision made on their acceptability. A record of this decision should be kept. A useful technique for letters is a rubber stamp, which can be applied to the reply letters from auditees; the stamp should leave enough space to allow both the recording of the decision and the noting of any further actions to be taken.

The results of the auditor's evaluation should always be provided to the auditee.

13.7   Further Actions

The "Further Action" line of area 5 is primarily intended for use when the action taken is unsatisfactory or action has not been taken by the effectivity point. However, it can also be used to note any special followup actions required. Area 5 also provides space for the auditor to sign and date the results he or she has recorded of the evaluation and any further action required. A copy of the request should then be provided to the addressee for information and possible further action.

13.8   Followup Actions

13.8.1   Introduction

The followup actions required depend on whether the corrective actions reported are found to be:

1. Acceptable
2. Not acceptable
3. No action reported

13.8.2   Acceptable

If the evaluation of the corrective action has not included a followup audit, then for external and internal quality audits alike one should be conducted at the earliest opportunity. This audit should evaluate all the functions connected with the problem concerned, including fringe or peripheral areas. Its results should be reported in the normal manner of reporting audit results.

If the corrective action taken is associated with a continuing internal quality audit procedure such as decision sampling, it can be of value in renewing confidence in the process to increase the degree of monitoring through larger samples or more frequent visits.

13.8.3   Unacceptable

If the evaluation determines the corrective action to be not acceptable, the seriousness of the remaining shortcomings should be assessed:

1. If they are relatively minor, another corrective action request, noting the shortcomings still present and requesting further action, should be made out and sent to the initial addressee.
2. If, on the other hand, the shortcomings are still major, a new corrective action request should be filled out and sent to the next level of management above the initial addressee. This should reference

the initial request and explain why the corrective action is unsatisfactory.

When dealing with unacceptable corrective action carried out by an outside organization where the initial request was sent to the designated quality representative, the followup request should go to the head of the organization. If the original request was sent to the head of the organization, any followup requests should be referred to the client for consultation.

13.8.4 No answer received

In regard to internal audits, if no answer is received to a request for corrective action by the effectivity point, a new request should be sent immediately to the next level of management outlining the problem and noting the lack of an answer.

In regard to external quality audits, the lack of an answer should be pointed out to the client in writing and guidance requested as to any further actions to be taken.

13.9 Corrective Action Log

13.9.1 Data entry

A corrective action log is a means of summarizing and collating the various corrective action requests initiated by the auditors. Ideally it should be entered into a computer data base so that routine sorts and analyses can be done by the computer. Data from both internal and external quality audits should be included, with external audit data being derived from:

1. External quality audits conducted on potential and actual suppliers

2. External quality audits conducted by potential and actual customers, certification agencies, etc.

I do not believe in logging the results of internal and external quality audits separately, as computers can readily segregate data in a variety of ways as part of any analysis. Moreover, analysis of the combined data might show interactions that would remain invisible if the files were kept separately.

The data is most effectively handled if entered in columnar form with columns for each of the following:

1. Type of audit—external or internal

2. Recipient of the request

  3. Location

  4. Date and reference number of the request

  5. Product or process involved

  6. Problem to be corrected

  7. Effectivity point requested for the corrective action

  8. Acceptability of the corrective action

  9. Actual effectivity when the corrective action was implemented

10. Name of auditor

11. Followup action

13.9.2 Data analysis

In analyzing this type of data, I recommend that each incident be considered on its own. I do not believe incidents should be converted into percentages that are considered in terms of some arbitrary acceptable quality level.

Each incident noted indicates a shortcoming in the process that results in one or all of the following types of problems:

1. Substandard product

2. Excess costs for correction

3. Delays due to correction

The analysis should collate these categories into Pareto distributions, simplifying the job of determining which occurrences should be corrected to reduce the incidence in each category to zero. Ideally a quality audit should confirm that all parties are making valid quality decisions with respect to their activities.

Alphanumeric sorting of particular columns in the data file brings together like entries. (Multitier alphanumeric sorts can also subdivide the main collation into collated subsets.) Counting of similar entries can then put the data into numeric form that will permit the rapid recognition of patterns. By including a program to develop Pareto distributions, the counts can sorted in order of magnitude.

For example, an initial alphanumeric sort of the "Problem" column will bring together all of the incidents relating to a particular problem. A secondary sort of the addressees will group the supervisors responsible for implementing the corrective action needed to solve that problem. This sort of addressees also indicates the scatter of the problem across the organization. If the problem is present in a number of areas, the cause may lie in the clarity of the requirements. This could lead to a clarification of the requirements.

This type of analysis can be particularly valuable in evaluating potential suppliers for conformance to a procurement quality system standard. A common problem area relating to a number of suppliers could indicate either a lack of clarity in the standard or an unusual interpretation by the auditors themselves. Both should be evaluated as part of the analysis.

Sorts of this nature are also helpful in evaluating:

1. Problems relating to addressees

2. Problems relating to product or processes

3. Product quality

4. Auditor problem patterns

5. Decision-making abilities

13.10 Conclusions

Corrective (quality audit) action requests are not documents intended to result in punitive action. They are intended to point out areas of activity where the required standard of performance is not being achieved. As addressees correct these shortcomings, their improved performance should be highlighted in reports so that the audit activity serves as a positive reinforcement promoting quality in the workplace.

In a company that operates an employee suggestion plan, quality auditors should *not* be eligible to receive compensation for any problems they note or improvements they suggest. Their job is to identify shortcomings and motivate corrective action. Thus including them on a suggestion plan eligibility list would serve to doubly compensate them for merely carrying out their job properly.

It should also be pointed out that with the breadth of modern quality programs, it is virtually impossible to identify a segment of an organization that is outside the scope of a quality audit program.

14

Reporting the Quality Audit

14.1 Introduction

The report showing the results of a quality audit is the raison d'être of that audit. It is the product of the audit activity, with all other audit activities simply being means to this end. And the quality audit report, whether for an external or an internal audit, must satisfy the needs of the customer, i.e., of the client. This principle applies regardless of the relationships between the client, the auditor, and the auditee. In all cases, decisions about the distribution of an audit report rest with the client.

In an external quality audit, the lead auditor has the responsibility to:

1. Determine the consensus of the audit team on the acceptability of the system reviewed

2. Make specific recommendations on behalf of the team to the client and auditee

The choice on acceptability is normally made from among four possibilities:

1. Unconditional approval

2. Conditional approval, i.e., the system is acceptable with minor improvements or corrective action

3. Conditional nonapproval, i.e., the system can be made acceptable with specific significant improvements or corrective action

4. Unconditional nonapproval, i.e., the system requires major redesign and development in order to be considered suitable for a further audit

The auditors know the information that has been derived from the audit, but it must be interpreted and presented in a manner that best satisfies the client's needs. Reports made to other areas of activity, e.g., to the auditee, to line departments, etc., will be at the discretion of the client. When issued, these will normally not be in the same format as the report to the client. The right to determine the acceptability of the quality system, process, or product being audited remains with the audit team, and is defined in the Terms of Reference agreed on by the client and the auditor.

Reports are presented in both verbal and documented forms. Clarity and accuracy are key requirements for every report, and the reporting format and language should be designed to readily convey the desired message to the recipient. Thus it may be necessary to use more than one type of report to satisfy differing needs. This could complicate the preparation of reports. However, with modern word processing equipment and personal computers (PCs), a variety of format presentations can be easily preset to minimize the labor involved in those preparations.

Since many of the reporting techniques used for external and internal quality audits are similar, this chapter presents a series of them, each of which can be used alone or in combination, as desired, to clearly present the audit results. The methods are illustrated in Figure 14.1, the function tree for reporting the quality audit.

Whenever practical, quality audit reports should be quantified (see Chapter 12). This permits the determination and illustration of performance trends. The need for quantification applies to both external and internal audits. However, it is particularly important for internal quality audits, where reaudits are normally much more frequent than for external ones, making trend analysis more practical. Quantified results can be presented in numeric or in graphic form. Both have their strengths and weaknesses.

14.2   Verbal Reports

14.2.1   Introduction

Verbal reports on the results of a quality audit fall into two fundamental categories:

1. Those presented in the course of the audit
2. The one presented at the debriefing meeting held at the conclusion of the audit

In both cases, it is essential that the data presented, and the conclusions or recommendations drawn from that data, agree completely

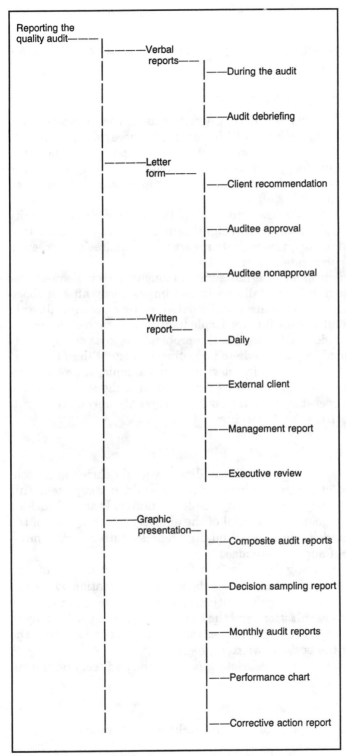

Figure 14.1 Function tree for reporting the quality audit.

with the corresponding information that will appear in the ultimate written report.

14.2.2 Verbal reports during the audit

Formal verbal reports will normally not be issued by auditors while an external quality audit is still in progress. There will of course be daily reviews of findings held between auditors and the lead auditor to determine if any changes or actions are required during the subsequent audit operations. But these discussions are not usually attended by any of the auditee's staff.

However, if the audit has found a serious breakdown in the quality system that involves safety or a significant change in plans, the lead auditor should make an immediate report to the auditee's representative so that appropriate action can be taken.

The feedback loop for internal audits is much shorter. Requests for corrective action will normally be issued immediately after a shortcoming is observed. These are usually delivered by an auditor directly to the individual responsible for implementing the corrective action. The two will undoubtedly discuss the problem. Reports or comments on findings should *not* be made to any other personnel than those responsible for the activities in question. For example, a problem observed on the shop floor should be discussed with the supervisor and *not* the actual operator, since the operator's activity may be the result of instructions from the supervisor.

14.2.3 Verbal debriefing reports

Traditionally for external quality audits, a verbal debriefing presentation is given at a meeting held when the audit is completed. This meeting is normally conducted by the lead auditor. It should include the senior management personnel of the auditee, with the head of the organization and its designated quality representative as the minimum number of auditee attendees.

Similarly, debriefing meetings are frequently held in conjunction with internal quality audits. But for these, the presentation by the auditors may be given to the executive and/or the lower management echelons of the organization, and they may also be held periodically to report on audit progress. Such meetings may precede or follow the preparation of the periodic written progress reports.

The purpose of a verbal debriefing is to let the auditee, and the client if present, know:

1. The conclusions of the audit team
2. The observations on which these conclusions are based

In regard to external quality audits, this will enable both sides to clarify their concerns before the final written report is issued. For example, the auditee may present additional data which did not come to light during the audit. This could affect the conclusions reached.

Prior to the meeting, the audit team must reach a consensus on:

1. Whether or not the quality system has met the requirements of the reference standard

2. If the quality system has met the requirements, whether there are any areas that require strengthening to assure it will continue meeting them

3. If it has not met the requirements, what areas require strengthening to achieve the desired standard of assurance

These conclusions must then be put in a point-by-point form so the lead auditor can make a clear and professional verbal presentation at the meeting.

The development of suitable visual aids will greatly strengthen the impact of the presentation and assist those attending the meeting to understand it. The visual aids should be as simple as possible to clearly carry the desired message. If slides or overhead transparencies are used, the data they contain must be in a format sufficiently large to be easily read by the attendees. Typewritten data should not be converted to slides unless the lettering is sufficiently large that it can be read from the projection masters at a distance of 6 to 8 ft.

For external quality audit debriefings, handouts of visual aids should *not* be provided at the meetings but instead should be included with the copies of the written report sent to the client. But for internal quality audit debriefings, visual aids *should* be provided to each attendee, along with a copy of the written report.

The actual presentation should include the following points:

1. A statement thanking the auditee for its courtesy and assistance, and acknowledging any special efforts put forward by the auditee's staff.

2. A statement about the purpose of the quality audit.

3. A statement about the reference standard or acceptance criteria used.

4. A clear statement about the acceptability of the auditee's quality system. (*Note:* Depending on the Terms of Reference for the auditors, this may reflect either the recommendations to the client or the decision of the auditors.)

5. If the quality system met the requirements, a statement about any particularly strong points, worthy of comment.

6. Also, if it met the requirements, a statement about any areas that require strengthening to assure it will continue meeting them.

7. Even if it did not meet the requirements, a statement about any particularly strong points, worthy of comment.

8. If it did not meet the requirements, an explanation of where it did not and what actions are necessary for it to meet them.

14.3   Reports in Letter Form

14.3.1   Introduction

Except as covering letters, reports in letter form primarily apply to external quality audits. Internal quality audit reports tend to consist of formal reports (see Sections 14.4 and 14.5), possibly accompanied by a covering letter and corrective action requests (Chapter 13). Letter reports tend to fall into the categories discussed below. To a large extent, the nature of a letter report depends on the freedom the audit team has been given in the formal Terms of Reference for the audit in regard to deciding whether or not the auditee's quality system meets the requirements of the reference standard and the team's freedom to report this decision to the auditee. The lead auditor should prepare and sign all correspondence.

14.3.2   Letter with recommendations to the client

In those cases where the approval/nonapproval decision remains with the client, all audit reports are made to that client. A report normally consists of a cover letter plus a detailed explanation of the audit's findings (the body of the report; see Section 14.4).

The content of the covering letter should include the following:

1. The name and address of the auditee

2. A statement of appreciation for the assistance given by the auditee

3. A statement about the reference standard used as the benchmark

4. A clear and concise statement of the audit team's recommendation as to the acceptability of the auditee's quality system

5. The title and reference number, if any, of the detailed report on the audit

In addition, some or all of the following points could be included:

1. An identification of any particular strong points in the quality program
2. An identification of the seriousness of any shortcomings observed, along with recommendations on the need for corrective action

14.3.3 Letters to the auditee on audit results

When approval/nonapproval decisions rest with the auditors, the lead auditor originates the necessary correspondence with the auditee. This is addressed to whoever is senior in the management of the auditee's organization or of the specific division being audited, e.g., CEO, division manager, plant manager, etc. A copy of all correspondence is also sent to the client.

If the decision is other than unconditional nonapproval, such letters should include the following points:

1. The name and address of the auditee.
2. A statement of appreciation for the assistance given by the auditee.
3. A statement about the reference standard used as the benchmark.
4. A clear and concise statement about the acceptability of the quality system.
5. If the decision is approval, a clear statement about the implications of approval, e.g., inclusion of the auditee in a qualified suppliers list (QSL), eligibility to tender for contracts, etc.
6. An identification of any particularly strong points in the quality program.
7. An identification of the seriousness of any improvements required or shortcomings observed, along with the need for corrective action. (These may be discussed in the letter along with the required effectivity points, or covered by attached corrective action requests. Writing as both an initiator and a recipient of external quality audit reports and letters in various forms, I prefer to see them discussed in attached request forms.)
8. Depending on the client/auditee relationship and client desires, an offer of quality engineering assistance to help implement corrective action.

The client's copy *only* should have the quality audit report attached to or referenced in it.

If the letter states unconditional nonapproval, items 5, 6, and 7 may be omitted. Depending on the client/auditee relationship and client's desires, quality engineering assistance can still be offered.

14.4 Written Reports

14.4.1 Daily reports

Every auditor, whether external or internal, should keep a daily record of his or her audit activities. This, in effect, becomes a daily professional diary. As a minimum this record should include: the auditor's name, dates and times of visits to the auditee, the reference standard being used, the observations made, and where and who made the observations. Such a diary can be of assistance in recalling points during a review of audit observations, in providing support data for the preparation of a report, in confirming when observations were made, in evaluating activities during a reaudit, etc.

The diary is usually kept in one or more of the following kinds of notebook, which have proven useful over years of audit activities:

1. A laboratory notebook, i.e., one with duplicate numbered pages, solidly bound but with the top copy of each page perforated for easy removal. The top copy goes to the lead auditor or audit supervisor for filing in the appropriate audit file. The carbon copy remains bound in the book for use by the auditor.

2. Weekly or monthly diaries with at least one page per day marked off in working hours to at least quarter-hour intervals. Those developed for logging activities in connection with improving the budgeting of time are excellent for audit purposes. As each diary is filled, it should be safely filed for use in report preparation and for future reference. These kinds of diaries are particularly useful for internal auditors, as they permit the supervisor to review how time is spent and if the audit visits are being suitably randomized.

14.4.2 External quality audit report for an external client

An external quality audit report prepared for an external client must satisfy the needs of the client and should provide:

1. The recommendations of the audit team with respect to the approval or nonapproval of the auditee's quality system

2. A comprehensive interpretation of the findings about each activity subject to the audit

This does not imply the inclusion of all the minutiae noted during the audit, but rather an interpretation of the effectiveness of the quality system plus details of each shortcoming noticed and the corrective action requested or recommended. Performance should be quantified, as far as possible. To ensure that the report is understood, the means used to quantify the data should be explained. (For methods of quantifying performance, see Chapter 12.) This report should be prepared by the lead auditor.

In preparing a quality audit report for an external client, consideration must be given to its appearance. It must be prepared in a clearly professional manner that will convey the image desired by the auditors and inspire confidence in their abilities and competence.

Normally, all reports should be issued in some form of binding or cover that identifies the contents as a quality audit report. This identification may be located on the outer cover or on a title page visible through a transparent front cover. A typical title page for this type of report is shown in Figure 14.2.

It is recommended that the title page be followed by a 100- to 200-word abstract or summary of the report. This summary should include:

1. The name of the reference standard used for the audit
2. The result of the audit, i.e., the degree of acceptability of the quality system
3. A brief review of the strengths and weaknesses of the various elements of the quality system

The main text of the report should follow on the next page. It should start with the report title and the author's name, followed by:

1. The location and dates of the audit
2. An acknowledgment of the assistance provided in the course of the audit, naming any outstanding contributors to its success
3. A statement about the purpose of the audit, including which reference standard was used
4. The detailed report itself

I believe that the detailed report should review the auditee's performance on a work-function-by-function basis and not on a quality-element-by-element basis. Each function should be reviewed in terms of the applicable elements of the quality system, as shown in Figure 14.3. (The planning guide illustrated in Figure 14.3 is based on procurement quality standard ISO 9001-1987.)

A.B.C. Manufacturing Co. Ltd.
Quality Audit Report

on

(enter the name of the auditee)

Reference Standard _____

Audit Team Members _____

_____

Date from _____ to _____

Prepared By _____ Date _____
(Lead Quality Auditor)

Approved By _____ Date _____
(Manager Quality Audit Department)

Figure 14.2   Quality audit report—title page.

While normally the reporting format will be left to the auditing organization, it must satisfy the needs of the client. In some cases, the client may prefer that the report be prepared in terms of the elements of the quality system. Quality elements are discussed in most procurement quality system standards, among them:

1. ISO 9001, 9002, and 9003 (issued in 1987)

2. ANSI/ASQC Q91, Q92, and Q93 (issued in 1987)

3. CAN/CSA Z299.1, Z299.2, Z299.3, and Z299.4 (issued in 1985)

4. BS 5750, parts 1, 2, and 3 (issued in 1987)

Figure 14.4 shows a planning guide for quality audit reports based on quality elements. Note that in these types of reports, work func-

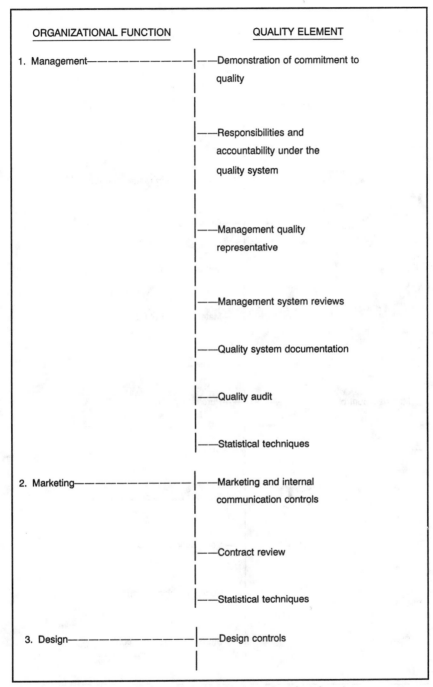

ORGANIZATIONAL FUNCTION   QUALITY ELEMENT

1. Management———————————|——Demonstration of commitment to
                                          quality

                                       |——Responsibilities and
                                          accountability under the
                                          quality system

                                       |——Management quality
                                          representative

                                       |——Management system reviews

                                       |——Quality system documentation

                                       |——Quality audit

                                       |——Statistical techniques

2. Marketing———————————|——Marketing and internal
                                          communication controls

                                       |——Contract review

                                       |——Statistical techniques

3. Design——————————————|——Design controls

Figure 14.3 Planning guide for a quality audit report based on functions.

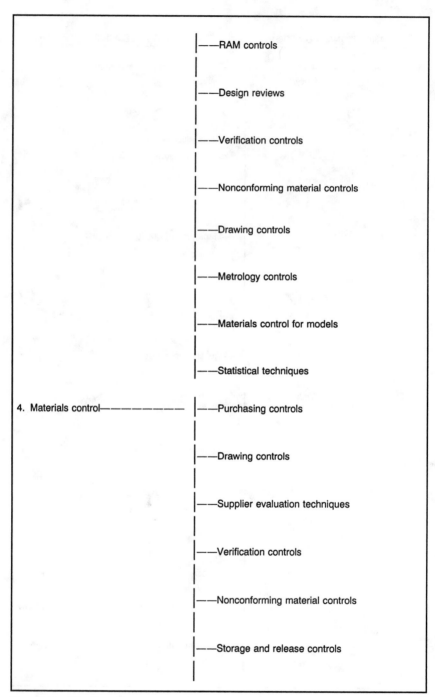

Figure 14.3   (*Continued*)

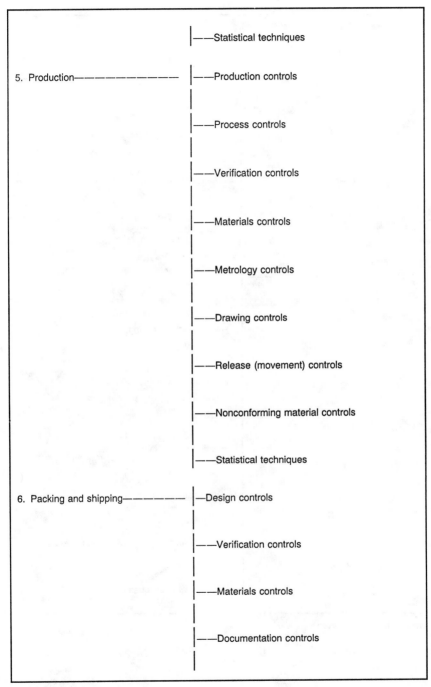

5. Production——————————
- ——Statistical techniques
- ——Production controls
- ——Process controls
- ——Verification controls
- ——Materials controls
- ——Metrology controls
- ——Drawing controls
- ——Release (movement) controls
- ——Nonconforming material controls
- ——Statistical techniques

6. Packing and shipping————
- —Design controls
- ——Verification controls
- ——Materials controls
- ——Documentation controls

Figure 14.3   (*Continued*)

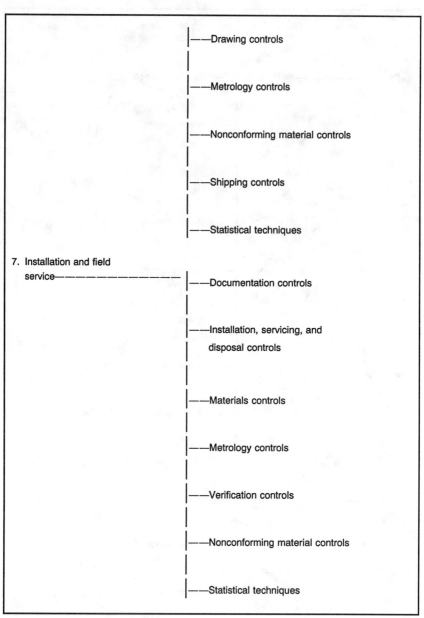

|——Drawing controls

|——Metrology controls

|——Nonconforming material controls

|——Shipping controls

|——Statistical techniques

7. Installation and field
   service———————————— |——Documentation controls

|——Installation, servicing, and
   disposal controls

|——Materials controls

|——Metrology controls

|——Verification controls

|——Nonconforming material controls

|——Statistical techniques

Figure 14.3   (*Continued*)

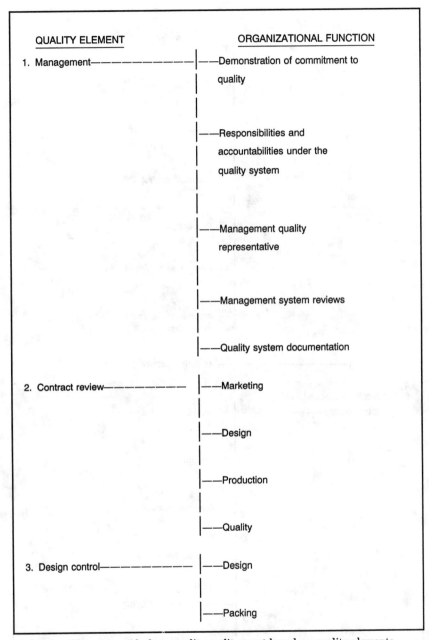

QUALITY ELEMENT   ORGANIZATIONAL FUNCTION

1. Management————————————|——Demonstration of commitment to
                         | quality
                         |
                         |
                         |——Responsibilities and
                         | accountabilities under the
                         | quality system
                         |
                         |
                         |——Management quality
                         | representative
                         |
                         |
                         |——Management system reviews
                         |
                         |——Quality system documentation
                         |
2. Contract review—————— |——Marketing
                         |
                         |——Design
                         |
                         |——Production
                         |
                         |——Quality
                         |
3. Design control——————— |——Design
                         |
                         |——Packing

Figure 14.4 Planning guide for a quality audit report based on quality elements.

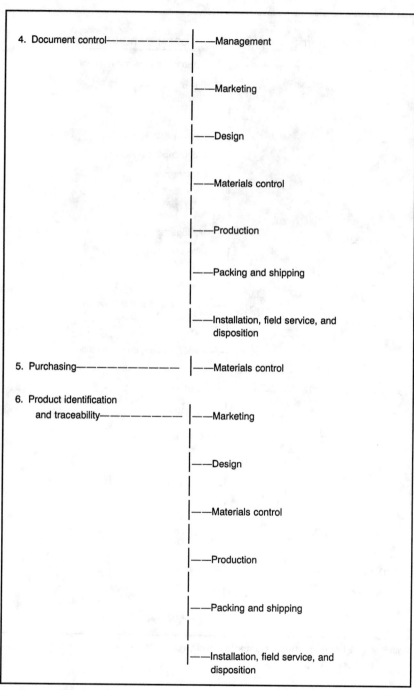

4. Document control————————— |——Management

|——Marketing

|——Design

|——Materials control

|——Production

|——Packing and shipping

|——Installation, field service, and disposition

5. Purchasing————————— |——Materials control

6. Product identification
   and traceability————————— |——Marketing

|——Design

|——Materials control

|——Production

|——Packing and shipping

|——Installation, field service, and disposition

Figure 14.4   *(Continued)*

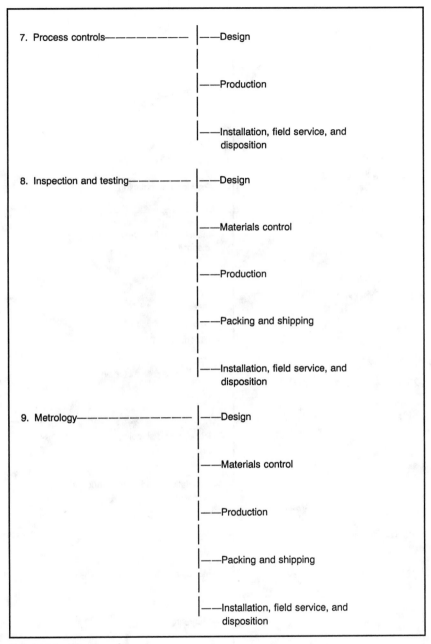

7. Process controls————————|——Design

|

|——Production

|

|——Installation, field service, and
        disposition

8. Inspection and testing————————|——Design

|

|——Materials control

|

|——Production

|

|——Packing and shipping

|

|——Installation, field service, and
        disposition

9. Metrology————————————|——Design

|

|——Materials control

|

|——Production

|

|——Packing and shipping

|

|——Installation, field service, and
        disposition

Figure 14.4   (*Continued*)

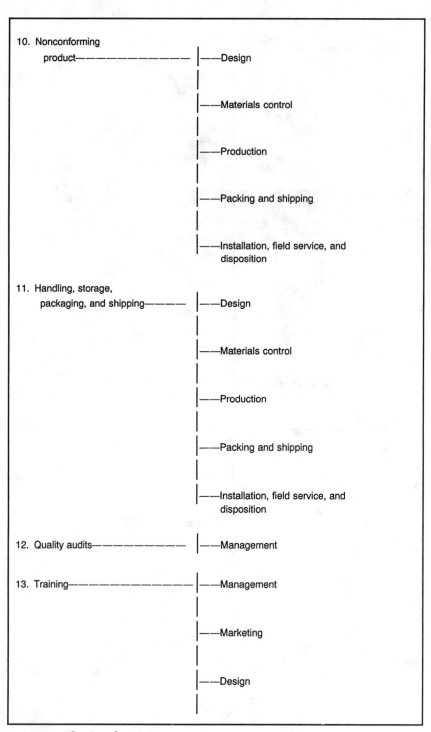

10. Nonconforming
    product——————————— |——Design

                                   |——Materials control

                                   |——Production

                                   |——Packing and shipping

                                   |——Installation, field service, and
                                      disposition

11. Handling, storage,
    packaging, and shipping————— |——Design

                                   |——Materials control

                                   |——Production

                                   |——Packing and shipping

                                   |——Installation, field service, and
                                      disposition

12. Quality audits————————— |——Management

13. Training————————————— |——Management

                                   |——Marketing

                                   |——Design

Figure 14.4 (*Continued*)

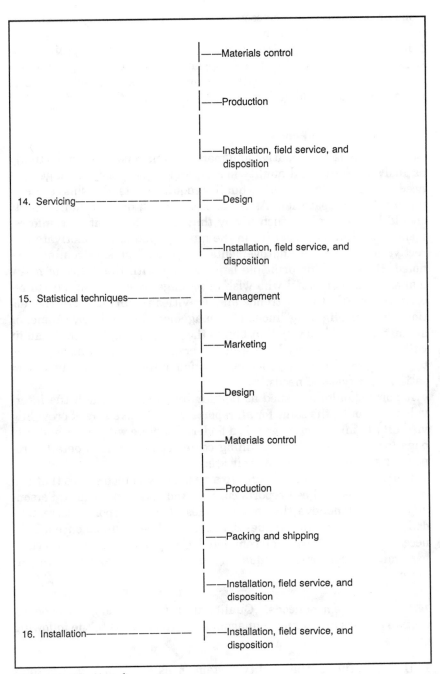

|——Materials control

|——Production

|——Installation, field service, and disposition

14. Servicing—————————— |——Design

|——Installation, field service, and disposition

15. Statistical techniques——————— |——Management

|——Marketing

|——Design

|——Materials control

|——Production

|——Packing and shipping

|——Installation, field service, and disposition

16. Installation——————————— |——Installation, field service, and disposition

Figure 14.4   *(Continued)*

tions or operations are classed as subsets of the quality system element(s) that apply to them.

If any corrective or improvement actions are required, details about them should be given in the appropriate section, including the reference numbers of any corrective action requests entailed. Copies of these requests should be indexed and attached to the report.

14.4.3 Management Report

14.4.3.1 General. Executive and management reports apply virtually exclusively to internal quality audits, and must satisfy management needs and attitudes as well as quality requirements. But this can complicate their preparation. As far as is practical, quality audit reports should be prepared in such a way that they offer positive reinforcement to acceptable activities in the quality system and motivate corrective action where improvements are needed or shortcomings are noted. However, this principle is in conflict with the desire of many management personnel who wish to manage by exception, i.e., to receive reports that show trends and activities requiring corrective action. This "firefighting" mode of management has to be overcome, or at least handled in a way that blunts the threat it implies, so the audit will not be regarded by lower-echelon employees merely as a prelude to punitive reactions. A judicious selection of reporting techniques can satisfy both types of needs.

Reports should be issued at regular intervals, although the intervals need not be the same for all reports. An effective way of providing visibility to middle managers is a folder, complete with cover and title page (see Figure 14.2), containing copies of the various detailed reports that are applicable to their areas.

The personal computer opens up a vast array of techniques that can readily be used to keep reports current and retain past data. Personally, I do not believe that paper copies of audit reports, correspondence, corrective action requests, etc., should be retained once all the necessary actions have been completed. Computer storage is less cumbersome and can provide equal or better accessibility to records, at lower cost.

14.4.3.2 Management needs. Quality audit reports should provide executive officers and senior management with answers to the following questions:

1. How effective is the quality system in each of our organization's major operating areas?
2. How can we improve the effectiveness of the quality system?

The reports must provide current information on work performance and assess longer-term trends as they become apparent. To keep the data current, reports should be issued at frequent intervals, e.g., monthly or quarterly.

Although an improvement program is a day-to-day, continuous activity, it requires planning and implementation over a longer period of time. Therefore, it is recommended that a comprehensive report be issued annually on the state of the quality system, following the outline given for external quality audit reports and the guide shown in Figure 14.3. This, in effect, is a management review report (see Section 14.4.4.3). But here the summary of improvements required also includes the plans for the on-going improvement program.

Lower- and middle-management personnel are more interested in the shorter-term performance and thus require:

1. Information on actual and potential problem areas that require corrective action

2. Reports that clearly show and interpret trends in their own performance and that of subordinates reporting to them

14.4.3.3 Corrective action request status reports. Immediate problem areas are identified to the applicable members of the management team through the corrective action requests issued by the auditors. This means that *not* receiving a corrective action request indicates the performance audited on a particular occasion was satisfactory. However, a monthly summary on the status of corrective action needed and requests issued can be of value to all managers and their superiors.

Typical monthly summary reports are shown in Figures 14.5 and 14.6. The two formats they illustrate indicate how easily report layouts can be prepared and modified on a PC. In either format, the number of vertical columns will depend on the size of the organization and the reproduction techniques available. Ideally the report should come off the computer printer ready to circulate. Each month the appropriate numbers should be entered in the squares, and the form reprinted and circulated again. In most organizations, the format shown in Figure 14.6 would be preferred since it provides full visibility regarding the requests issued, and whether they have been completed or not completed. However, the format shown in Figure 14.5 can readily be adapted to do so too.

Additional pages can be attached to these reports, listing outstanding corrective action requests grouped by supervisor responsibility.

14.4.3.4 Internal quality audit—status reports. An internal quality audit often tends to be a continuous process covering some of the quality

A.B.C. Manufacturing Co. Ltd.
Monthly Summary
Corrective Action Requests

Department _____

Activity/ Month	Shop A		Shop B		Shop C		Shop D		Shop E	
	Issued	Completed	Issued	Completed	Issued	Completed	Issued	Completed	Issued	Completed
January										
February										
March										
April										
December										

Audit Department _____ Date _____

▨ Idle Area

Figure 14.5 Corrective action requests—monthly summary form.

system elements in different operational areas on a weekly basis. From a practical auditing program and reporting cycle point of view, an effective internal quality audit program should cover all the elements of the quality system in each operational area at least once in a quarter, which nominally covers a 13-week period. To achieve this, more than one auditor may be involved in the audits of a given area. Therefore, it is essential that each auditor be aware of the factors covered by the others and the results of their audit activities. An internal quality audit status report (Figure 14.7) provides a ready method of preventing duplication or gaps in the audits, as well as a way of reporting the findings of each auditor.

This type of report has been used successfully with 10-, 13-, and 26-column formats. The smaller sizes are recommended for use with report preparation on a PC. The 26-column format is suitable for preprinted

A.B.C. Manufacturing Co. Ltd.
Monthly Status Report
Corrective Action Requests

Department _____

Activity	Shop A			Shop B				Shop E		
	Issued	Comp- leted	n/c	Issued	Comp- leted	n/c		Issued	Comp- leted	n/c
January										
February										
March										
April										
December										

Audit Department _____ Date _____

n/c – Not Completed

◩ Idle Area   (Outstanding Requests)

Figure 14.6   Corrective action requests—monthly status report form.

8½- by 11-in forms, using colors instead of the cross-hatching identifications shown in Figure 14.7. The time periods can represent daily, weekly, or monthly intervals depending on the application.

In preparing this type of report for use as an on-going record, it must first be decided whether to keep it in terms of quality system elements (see Figure 14.4) or the functional or operational elements of the organization (see Figure 14.3). Then each primary element of the approach selected is entered on the "Function/Location" line near the top of the form, and all the applicable subelements are entered line by line in the vertical column headed "subelements." (This of course means that a separate sheet is required for each quality system element or work function or operation being audited.)

Figure 14.3 provides a breakdown of applicable work functions or operations, and Figure 14.4 of quality system elements. Referring to

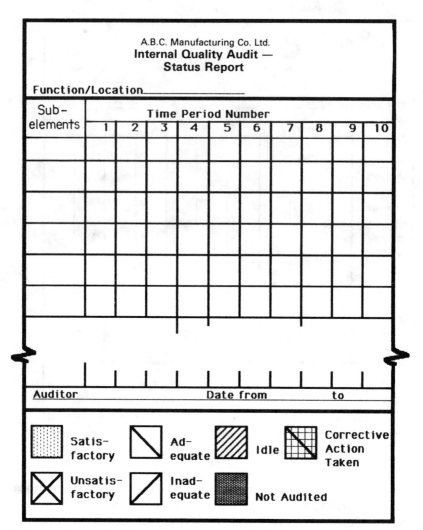

Figure 14.7   Internal quality audit—status report form.

them when preparing the appropriate status report will help ensure that no area is overlooked.

Let us assume a report is to be made about a manufacturing organization in terms of the elements of its quality system. Each of the following topics would then have to be covered (each on its own sheet):

1. The presence of documented procedures showing what is required to achieve the objectives of the quality system

2. Demonstrated conformance to the procedures

3. That shop methods data are being used for the manufacturing activities, defining how the product is produced

4. That the latest drawings, specifications, and other product documentation are available and being used

5. That acceptance criteria for all characteristics are defined and being used

6. That verification procedures are defined and being followed

7. That all instrumentation, gauges, fixtures, etc., are controlled and calibrated in accordance with the quality system requirements

8. That process capability and control records are available and being used

9. That good housekeeping procedures are in effect

10. That good safety practices are being followed

Each functional or operational element of the organization to which the quality system element applies would then be listed in the "subelements" column. Each of these should be broken down at least to the lowest level of supervision, although at times it may be necessary to use even smaller elements.

At the beginning of each time period, normally a week, the cross-hatching used to identify idle areas should be added to the computer master and the form printed out anew. This printout, showing the results from the previous periods and the idle areas, then becomes the working master form for that week.

All internal quality auditors should use a single sheaf of records to log their findings. Each auditor enters the results of individual audits on the working master and the computer master using the relevant symbols, e.g., diagonals, cross-hatching, etc., to show the overall results. As the audits continue, patterns will become apparent that allow individual work performances to be assessed.

In practice, a manual version of this format was used in a major manufacturing area for several years. The record was borrowed by the manager of manufacturing each week to use in the weekly production review.

With the use of a PC, copies can easily be provided to management as part of the periodic report on performance.

14.4.3.5 Decision sampling reports. The internal quality audit status report can be used for reporting performance with respect to decision sampling. In this application, a line is provided for each individual making quality decisions and subject to sampling. The intervals decision sampling intervals are frequently daily when dealing with verification personnel, i.e., inspectors, testers, etc. Performance status can be indicated using markings similar to those shown in Figure 14.7.

14.4.3.6 Quality system performance reports. When the full organization has been subjected to a complete internal quality audit, a report should be issued comparing the performances in various areas. This report should then be repeated each time the audit cycle is completed. For a 13-week cycle, the report would be quarterly.

The performance level of each supervisor in the organization is calculated using one of the techniques outlined in Chapter 12. The method and confidence level used should be the same for each calculation so that comparisons between performances are on a common basis. Once the individual performance levels have been calculated, the organizational average $(\overline{X})$ and the control limits (± 3 s) can be calculated.

A bar graph is then plotted showing each supervisor's performance. Horizontal lines are added to show the average $(\overline{X})$ and the control limits falling within the graph scale. An example of this type of report is shown in Figure 14.8. In an actual report, the confidence level in the

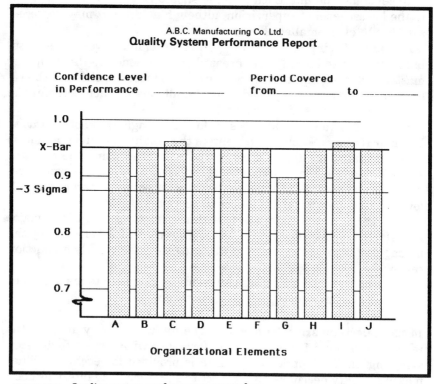

Figure 14.8   Quality system performance report form.

performance and the period covered by the report should be filled in.

This type of report clearly shows comparative performances and the areas requiring improvement.

14.4.3.7 Performance trends. Performance trends can be plotted for each operational element as well as for the overall program. Either of two normal process control charting techniques can be used for this:

1. Plotting the performance levels as the $\overline{X}$ values on an $\overline{X}$ and R control chart

2. Using the difference between the actual performance level and an arbitrarily selected normal value to plot a CuSum control chart

Normally, control limits need not be calculated for these charts since they are intended to show trends and not provide absolute measurements.

I prefer the use of the CuSum technique where improvement or degradation of performance is clearly shown by the slope of the graph. In selecting the norm for this type of chart, one can use either the ideal objective of 1.0 denoting perfect performance, a value approached, but never reached in calculating the statistical performance level, or some selected lower value. However, I personally do not recommend the use of 1.0 as the norm, since it will always result in a negative difference and hence a downward slope on the graph, which implies negative rather than positive reinforcement for the individuals whose performance is being monitored. Instead, I recommend the use of the initial performance level as the norm. Then improved performance will result in a positive difference and an upward slope to the graph, and a degradation in performance will result in a negative difference and a downward slope. A quality system is targeted toward constant improvement in performance; hence a successful system will result in a CuSum control chart with an upward trend, which implies positive reinforcement for the individuals whose performance is being monitored.

14.4.3.8 Periodic audit reports. In some organizations, an internal quality audit is carried out on a function-by-function basis, i.e., each organizational element is audited in regard to all aspects of the quality system during a single audit. In this approach, the interactions between the various elements of the quality system related to that organizational element are evaluated. However, the approach is not particularly sensitive to interactions between organizational elements. Audits carried out on this basis should be planned to cover all the operational elements within a given period of time, say once each quarter. And since some audits should be completed each month, a monthly summary report should be issued.

If, on the other hand, the internal audit is carried out on a continuous basis as discussed in Section 14.4.3.4, the periodic audit reports are normally issued quarterly. (And thus the titles of Figures 14.9, 14.10, and 14.11 should all read "quarterly," and the left-hand columns should show quarters instead of months.) With this approach, a single sheet can readily cope with several years of data, thus helping to make any long-term patterns more apparent.

A typical monthly report form for a small organization is shown in Figure 14.9. Each vertical column is allocated to a particular element of the organization. The number of vertical columns will depend on the number of elements concerned.

In a larger organization, the executive and senior management levels will undoubtedly prefer a report that breaks down the organizational activities under more major headings, as shown in Figure 14.10. In such cases, the performances of the lower-echelon employees are grouped according to single members of the senior management team, so they can easily see the combined performance of those reporting them. A more detailed breakdown of each group would then be shown on a form similar to that illustrated in Figure 14.11. These more detailed reports can be attached to the executive summary sheet (Figure 14.10) and provided to all senior members of the management team, or just provided to the chief executive and the responsible manager. The example shown in Figure 14.11 covers Materials Control.

The results of the audit and subsequent corrective actions should be illustrated by use of the appropriate graphics. Such charts are readily producible on a PC. However, they can also be prepared manually by the use of rubber stamps applied to a master print, which can then be reproduced for circulation.

Both types of reports—for a small or a large organization—should be supported by a brief summary of the audit findings for each area noted on the forms, similar to the "Summary" section described for an external quality audit report (see Section 14.4.2).

14.4.4 Executive report

14.4.4.1 General. There are two types of internal quality audit reports required at the executive level. The first, discussed in Section 14.4.3.8, provides periodic information on the status of the quality system. The second provides the information necessary for a "management review" of the adequacy of the system to determine any changes in policies, philosophy, directives, or methods needed to improve it.

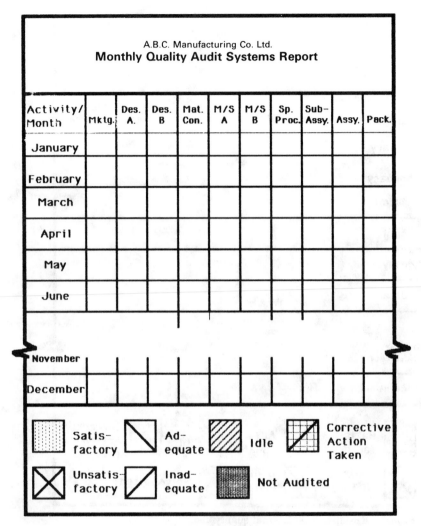

A.B.C. Manufacturing Co. Ltd.
Monthly Quality Audit Systems Report

Activity/ Month	Mktg.	Des. A.	Des. B	Mat. Con.	M/S A	M/S B	Sp. Proc.	Sub- Assy.	Assy.	Pack.
January										
February										
March										
April										
May										
June										
November										
December										

- Satis-factory
- Ad-equate
- Idle
- Corrective Action Taken
- Unsatis-factory
- Inad-equate
- Not Audited

Figure 14.9 Monthly quality audit status report form—small organization.

14.4.4.2 Executive status reports. Most executives desire summaries that show the status of the overall quality system with respect to their organization's defined quality policies and the supporting procedures intended to implement the policies. They will only desire detailed information on the system when they are reviewing potential improvements.

This type of information can be readily conveyed by periodic quality audit status reports (see Section 14.4.3.8). The symbols used on these forms provide ready evidence as to the status of the quality system in

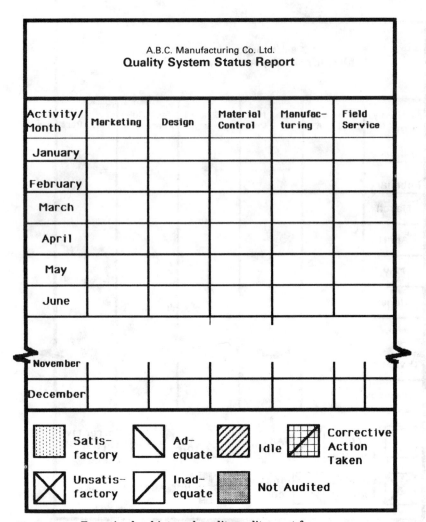

Figure 14.10   Executive-level internal quality audit report form.

relation to various work functions. The forms should be accompanied by additional sheets containing one-paragraph summaries of any areas requiring corrective action. They can also be augmented by selected performance trend charts (see Section 14.4.3.7).

Reports of this nature permit the executive to raise questions with the responsible managers regarding their adherence to the quality system. The type of status report shown in Figure 14.10 is based on the principle that each line manager is responsible and accountable for the quality of his or her function, in terms of both production and verification activities. When this is the case, the quality function be-

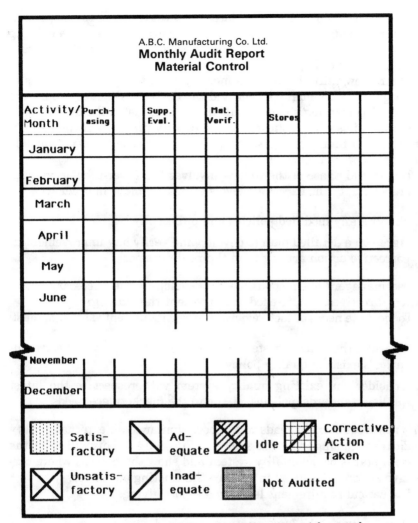

A.B.C. Manufacturing Co. Ltd.
Monthly Audit Report
Material Control

Activity/ Month	Purch- asing		Supp. Eval.		Mat. Verif.		Stores			
January										
February										
March										
April										
May										
June										
November										
December										

▦ Satis-factory	◻ Ad-equate	▨ Idle	▦ Corrective Action Taken		
⊠ Unsatis-factory	◪ Inad-equate	▨ Not Audited			

Figure 14.11 Monthly quality audit status report form—materials control.

comes a resource center responsible for assisting in the documentation and development of the quality system, and the department responsible for auditing that system for conformance.

14.4.4.3 Management review reports. Good management practice, as well as certain procurement quality system standards issued by ISO, ASQC, CSA, BSI, etc., require a periodic "management review" of the quality system. This is to ensure that the system reflects the latest desires, policies, philosophy, directives, etc., of the organization and the latest requirements of the marketplace. For such a review to be effective, the quality audit report must provide clear evaluations of the existing system against the new requirements. It must also address the areas requiring change or improvement.

The review initially involves meeting with the executive levels of the organization's management and the designated quality representative. This meeting should result in a consensus on the desired policies and philosophy applicable to the quality program. If the deliberations result in any changes, the quality representative is responsible for documenting them.

The second phase of the review involves the lower-echelon managers reporting to the executive levels of management in order to:

1. Finalize any modified policies and philosophy
2. Agree on a detailed plan to develop and verify any improvement or corrective action arising from the modifications

These managers are responsible for developing the necessary procedures, directives, etc., needed to implement the plan agreed on.

To meet the needs of such reviews, the reports presented at them must:

1. Consider the quality system as a whole with respect to the organization's existing quality policy
2. Consider the existing quality system with respect to the latest quality standards, policies, etc., and the marketplace needs

The first requirement leads to a report similar to an external quality audit report, using the guide shown in Figure 14.3. This will cover the present status of the quality system and show those areas where corrective action is required to achieve the existing objectives.

The second requirement leads to a report giving:

1. A summary of the improvements required.
2. A detailed analysis, by operating functions, of the overall system with respect to the latest quality needs. For example, this could cover such factors as the need to change from a philosophy of correction to one of prevention, the need to apply the quality system to areas not yet functioning within it (clerical, accounting, etc.), and so forth.

Both types of reports should make use of the graphical aids necessary to ensure their clarity to meeting participants.

Distribution of these reports in whole or in part depends on the desires of the executive level of management. The minimum distribution would be:

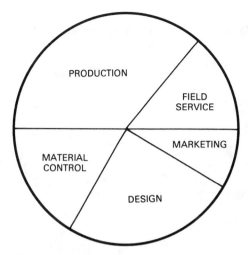

Figure 14.12   Corrective action "pie chart."

1. Whole report to executives
2. Summaries and functional sections to the responsible managers

14.5   Graphics Presentations

With the current availability of graphics programs for use with PCs, the only limiting factor on their use is the creativity of those preparing the reports. Examples of the use of graphics programs have been given in the above paragraphs. Other techniques include "pie charts" showing, for example, the distribution of corrective action requests issued or outstanding for each operational element. See Figure 14.12.

14.6   Conclusions

The report is the raison d'être for a quality audit, and thus care must be taken to ensure that it carries the desired message to the intended reader. Judicious selection of text and graphic techniques should ensure that any such report satisfies the most critical of recipients. This chapter has given examples of reporting techniques which have been used successfully at various times by the author. Quality audit reports will continually be changing in their formats and approaches as new individuals are involved in preparing and using them. As techniques improve, e.g., by means of the use of new generations of personal computers and their programs, so improved and more effective reports should result.

Quality audits and quality audit reports should be constantly improving as much as any other function or operation within an organization. In fact, they should lead all other creative activities in this, by setting an example of an open-minded approach.

Measuring and Improving the Effectiveness of the Quality Audit

15.1 Introduction

The subject matter of this chapter has somewhat different connotations for the management of external or internal quality audits, but there are also certain areas of similarity. Examining Figure 15.1, the function tree for measuring and improving the effectiveness of the quality audit, shows the differences to lie largely in the area of determining the effectiveness of the different types of audit, whereas the similarities lie more in the areas of managerial responsibility and the ways that can be used to improve audit operations.

15.2 Managerial Responsibility

The manager of quality audit activities, as with all managers in an organization, has the responsibility for continuously improving the effectiveness of the process under his or her control. In some ways this is a more abstract process than many of the others carried out within an organization. Therefore it requires more ingenuity to initiate and maintain an improvement program in this field.

The first essential point in any improvement program is to know the starting point or present status of performance in quantitative or measurable terms. Without this reference point, one cannot be certain that any changes introduced are resulting in an improvement rather than a degradation of performance effectiveness.

Therefore, the first step in regard to improving the effectiveness of a quality audit program is to determine the present status of that pro-

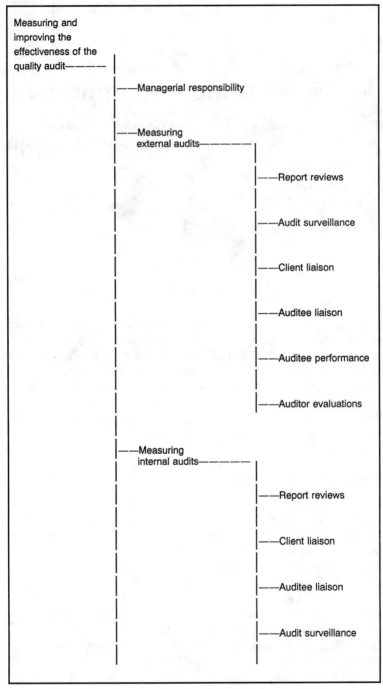

Figure 15.1 Function tree for measuring and improving the effectiveness of a quality audit.

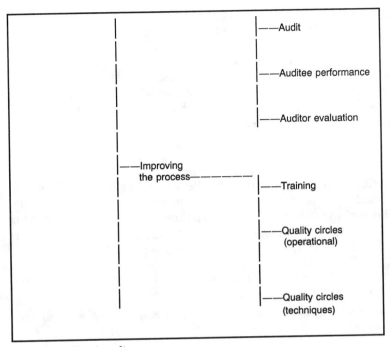

Figure 15.1 *(Continued)*

gram in measurable terms. Once this has been done, an improvement program can be developed and introduced and comparisons made with the starting or reference performance. The improvement program should be a continuing element of the quality audit program. The manager of quality audit activities must be continuously looking for ways to improve the audit process and encouraging the auditors along the same lines.

15.3 Measuring the Effectiveness of External Quality Audits

15.3.1 Introduction

Measuring the effectiveness of the quality audit process depends to a large extent on the interrelationships between the client, the auditee, and the manager of audit activities. Aside from any organizational relationships, the interrelationships referred to here include such factors as professional respect and respect for different technical know-how, as well as the psychological interactions that occur between the three parties.

It must, of course, be recognized that the company or organizational relationships that exist will depend on the type of quality audit concerned (see Chapters 2 and 3). And these relationships in turn will have some bearing on the detailed audit techniques used (although not on the general philosophy guiding the assessment). They may also, at times, cause the assessment of the effectiveness of external quality audits to be more judgmental than the assessment of the effectiveness of internal audits.

15.3.2 Report reviews

Each quality audit report issued, regardless of the approval chain, should be analyzed and reviewed by the audit manager. This review should cover the working papers used by the auditors and any form of reporting used to aid in the development of the final report. The effectiveness of the report should be analyzed in terms of the following factors:

1. Does the report reflect the professional and technical image desired for the quality audit activities?

2. Have all the observations reported been related to the applicable sections of the reference standard?

3. Is the report written in clear and unambiguous language?

4. Does the language convey the findings to the intended reader, i.e., to the client and the auditee?

5. Does the report show any biases by individual auditors through undue concentration on particular processes or elements of the quality system?

6. Do the findings, requests for action, and recommendations fall within the terms of reference of the quality audit?

Besides the manager forming his or her own assessment of the report's effectiveness, a thorough analysis requires discussions with the client and the auditee. And by collating the data from a number of different audits and auditors, it is possible to assess the effectiveness of the audit teams and auditors in reporting their findings in a clear and objective manner.

15.3.3 Audit surveillance

Periodically the audit manager should carry out a surveillance of the quality audit team in action. Surveillance permits the manager to observe and assess the techniques and effectiveness of the audit team, the team leader, and the individual auditors under the stresses of actual audit conditions. This surveillance can take a number of different forms, depending on the size of the audit and its activity load:

1. Actual surveillance of an audit in progress, i.e., attending and observing various audit activities but *not* participating in them. This is, in effect, an audit of the audit activities. Great care must be taken with this technique not to participate in, or to influence, the planning, implementation, or analyses of the findings in any way.

2. Participation in the quality audit as the team leader, which allows the assessment of the various members of the team under actual audit conditions. In this instance, care must be taken to ensure the auditors make the decisions they are required to make without feeling any undue influence from the manager.

3. Participation in the quality audit as a member of the team, but not the team leader, which permits the assessment of team leaders under the stress of actual audit conditions. Here once again it is important that the manager make only those decisions pertinent to the level of his or her participation in the audit activity. The manager should *not* interfere with the team leader's carrying out his or her normal responsibilities or actions.

15.3.4 Client liaison

The client, as the customer of the quality audit, has the real facts on whether the quality audit is satisfying his or her needs. Depending on the relationship between the client and the auditee and the type of quality audit, the client(s) may be single or multiple. An example of a type of external quality audit having a single client is the evaluation of potential suppliers for inclusion on a qualified supplier list to be used by the procurement department of an organization. Examples of the types of external quality audits having multiple clients are: audits undertaken to determine the suitability of a quality system in meeting the needs of organizational management, or those undertaken at the requests of different organizations desiring the registration of their quality systems as acceptable to a particular agency or certification board.

The degree of satisfaction of the client(s) can be determined by liaison meetings held either periodically during the audit or at its end, or by questionnaires filled out by the client(s) after the audit is completed. A typical questionnaire is shown in Figure 15.2.

If meetings or interviews are used as the liaison technique, working papers similar to the questionnaires should be prepared to ensure that all aspects of the client's needs are being considered and satisfaction graded.

A.B.C. Manufacturing Co. Ltd.
Client Evaluation of the External Quality Audit

(Please rate each question from 1 to 5 with 5 implying excellent satisfaction.)

Auditee_____

1.0 How does the overall Quality Audit Report satisfy your requirements?

    ☐ _ ☐ _ ☐ _ ☐ _ ☐

    1   2   3   4   5

What areas require improvement _____

_____

2.0 Is the text sufficiently clear and understandable in its language?

    ☐ _ ☐ _ ☐ _ ☐ _ ☐

    1   2   3   4   5

What areas require improvement _____

_____

3.0 Is the text sufficiently clear in its presentation?

    ☐ _ ☐ _ ☐ _ ☐ _ ☐

    1   2   3   4   5

What areas require improvement _____

_____

Figure 15.2 Typical quality audit evaluation questionnaire for clients.

Once the data is available, its collation and analysis follows steps similar to those outlined in Chapter 12. Quantitative statements should be used wherever possible. Since, in general, "quantitative" in this context means statistical measurements, a confidence level must always be decided on and stated in any reports generated. As the number of clients increases, confidence levels can be increased and more subtle patterns may be revealed.

15.3.5 Auditee liaison

Although the client is the customer of a quality audit, the audit's effectiveness depends to a large degree on the cooperation of the auditee. Therefore, it is essential that the auditee be satisfied that the audit has been conducted professionally and reported objectively, providing a true and accurate evaluation of all the activities audited. The auditee must also be satisfied that the audit procedures caused the minimal interference possible to normal work operations.

From an external quality audit point of view, each auditee is a separate and distinct entity that must be considered individually. Liaison with an auditee thus becomes a one-on-one situation. This may involve a single meeting once all the audit activities have been completed, including any requested corrective action, or it can mean a questionnaire left with the auditee at the time of the debriefing meeting. A typical questionnaire for auditees is shown in Figure 15.3. As with client liaison, if a meeting is held, working papers covering all the points a questionnaire would deal with should form the agenda.

As noted in regard to data gleaned from clients, data from auditees can be analyzed using the techniques outlined in Chapter 12. Again the information sought will be largely statistical in nature and will involve confidence levels. As data is received from a broader cross section of auditees, more subtle patterns may develop that require further investigation.

15.3.6 Auditee performance

Where an external quality audit has been conducted to determine the suitability of a potential supplier and contracts have resulted, the findings of the audit can be compared with the actual performance of that supplier in fulfilling the contracts. This data can be extracted from the verification records on goods and services received from outside suppliers. It may be derived through source surveillance or the examination of records, receiving inspection records, data on production problems resulting from supplier falldown, etc. Supplier shortcomings found in actual practice may reveal shortcomings overlooked by the audit team or a deterioration in the performance of that supplier. Either can result in an examination of the techniques, procedures, etc., used by the audit team to determine what steps can be taken to improve the effectiveness of the activity. This type of problem solving may result in corrective actions being taken by both the supplier and the audit team.

15.3.7 Evaluation by the auditors

Once all of the activities associated with a given audit have been completed, a postmortem meeting of the audit team members should be held. The purpose of this meeting is to identify any audit implementation problems encountered and develop techniques for reducing the

---

A.B.C. Manufacturing Co. Ltd.
Auditee Evaluation of the External Quality Audit

(Please rate each question from 1 to 5 with 5 implying excellent satisfaction.)

Auditee_____

1.0 Were the pre-Audit arrangements carried out to your satisfaction?

☐ _ ☐ _ ☐ _ ☐ _ ☐
1   2   3   4   5

What areas require improvement _____

_____

2.0 Did the Briefing Session answer all of your queries?

☐ _ ☐ _ ☐ _ ☐ _ ☐
1   2   3   4   5

What areas require improvement _____

_____

3.0 Were you satisfied that interference with your productive activities was kept to a minimum?

☐ _ ☐ _ ☐ _ ☐ _ ☐
1   2   3   4   5

What areas require improvement _____

_____

4.0 Were you satisfied with the interface relationships between the Auditors and your personnel?

☐ _ ☐ _ ☐ _ ☐ _ ☐
1   2   3   4   5

What areas require improvement _____

_____

Figure 15.3 Typical quality audit evaluation questionnaire for auditees.

probabilities of their recurrence. The meeting should review the various problems encountered by the audit team in carrying out the particular audit. The observations made on the auditee's activities should *not* be reviewed, except insofar as an audit method or technique might have contributed to a questionable observation or nearly miss observing a particular problematic condition.

As implementation difficulties are identified they should be listed. I do not believe any attempt should be made to develop solutions to particular problems until the team is certain that all the problems have been identified and listed.

The detailed list should then be reviewed by the team members to determine if any of the items noted can be grouped into sets of similar problems. The final list should then be prioritized, Pareto distribution being the most useful tool for this. In cases where different problems have the same number of occurrences, the sequence of tackling them should result from a group consensus or, if that is not possible, should be decided by majority opinion.

Quality circle techniques can be used for the postmortem meeting. See Sections 15.5.2 and 15.5.3, where the use of quality circles is discussed in a broader context.

15.4 Measuring the Effectiveness of Internal Quality Audits

15.4.1 Introduction

Internal quality audits provide much more adequate data for evaluating the effectiveness of the audit procedures than do external quality audits. As all parties involved work for a common organization, there should be a freer exchange of ideas about, irritations over, and other reactions to the audit process. Of course compartmentalization of work functions can hinder these discussions. But this merely provides a further challenge to the audit team to demonstrate the overall benefits of the audit and point out how its aim is constructive in nature, intended to assist the different work functions rather than gather evidence that could lead to punitive actions.

Although internal quality audits have a single client, i.e., the person initiating or authorizing their implementation, there are normally several "customers" or users of the audit data output. All of these "customers" work for the same organization. This is advantageous to an audit improvement program because all of them are in a good position to actively contribute to it.

Similarly, auditee input to an improvement program is also diverse and close at hand. Instead of only a senior member of the organization being involved, as is the case for external audits, all of the supervisors or managers in charge of the audited areas can participate in the evaluation, supplying their input on the methods, the techniques, and even the results of the audit.

Internal audits are normally conducted more frequently than external ones, and thus data can be derived more quickly on existing and improved techniques. As all three parties—client, auditee, and auditor—are part of the same organization, planned experiments can be introduced to determine the most effective techniques.

For these reasons, internal quality audits more readily lend themselves to a program of measuring and improving audit techniques and the skills of the auditors.

15.4.2 Report reviews

Internal quality audit reports fall into two general categories:

1. Periodic performance reports, normally issued over the signature of the audit manager
2. Requests for corrective action issued by the auditors in the course of their activities

The effectiveness of the first type will be made apparent during the presentation of such reports to the management team concerned and any subsequent client liaison activities (see Section 15.4.3).

Reviewing the effectiveness of the second type should be broken down into reviewing the clarity and effectiveness of the requests themselves, and reviewing the effectiveness of the proposed improvements or corrective actions.

Periodically, requests for corrective action should be reviewed by the audit manager. These in-process reviews will show if auditors have interpreted their observations correctly in order to identify the underlying problems and who is responsible for implementing the corrective action. The periodicity of these reviews will depend on the number of requests that must be reviewed. If they are numerous, it may be necessary to review only samplings of the requests made by each auditor. But also note that if corrective action requests are this numerous, it probably indicates a major breakdown in the quality system being audited that requires urgent action.

15.4.3 Client liaison

With internal quality audits the client is the executive who authorizes the quality audit activities. However, each member of the management team reporting to that executive is a "customer" of the audit who must be provided with the data derived from it in a form that will most effectively convey the information he or she requires.

How satisfied these customers and the client are with different aspects of the audit can be determined by contacting them individually or by holding a general meeting to review the effectiveness of the audit program. A meeting of this nature may, or may not, include the organization's chief executive officer. Note that most procurement quality system standards require a "management review," and at least one meeting per year should be held including the client and all the customers to review comprehensively the entire quality system. However, this review should be separate from any meetings held to identify problems or dissatisfaction with the audit process itself.

The agenda for the discussion should follow the outlines indicated by the questions shown in Figure 15.4. This request for information from executives and managers regarding their reactions to the audit should be sent to all attendees prior to the meeting so they are aware

---

A.B.C. Manufacturing Co. Ltd.
Executive Evaluation of the Internal Quality Audit

(Please rate each question from 1 to 5 with 5 implying excellent satisfaction.)

Executive_____

1.0 How does the overall Quality Audit Report satisfy your

requirements?

☐ _ ☐ _ ☐ _ ☐ _ ☐
1   2   3   4   5

What areas require improvement _____

_____

2.0 Is the text sufficiently clear and understandable in its

language?

☐ _ ☐ _ ☐ _ ☐ _ ☐
1   2   3   4   5

What areas require improvement _____

_____

3.0 Is the text sufficiently clear in its presentation?

☐ _ ☐ _ ☐ _ ☐ _ ☐
1   2   3   4   5

What areas require improvement _____

Figure 15.4 Typical internal executive and managerial information request.

of the topics to be covered. The questionnaire can also be used in lieu of a meeting, although some effectiveness of the reporting system will be lost.

Since the meeting is intended to identify elements of the audit system requiring improvement, all attendees should be encouraged to speak frankly and honestly. The audit manager should not be defensive about existing audit techniques, but must be willing to listen to and take note of the criticisms that arise. Solutions to problems may be suggested by attendees during the meeting, but their adoption should be left until the full data package on problems is available. Deciding on solutions too early might lead to interactions between different problem areas being overlooked.

Four fundamental questions must be considered in finalizing solutions for any problem:

1. What does the problem involve?

2. What does the problem not involve?

3. How does the problem or intended solution interact with other identified problems?

4. Does the intended solution adversely interact with other elements of the audit system?

15.4.4 Auditee liaison

With internal quality audits, it is relatively convenient to bring together the supervisors of the audited areas. In a large organization, bringing them together at a single meeting could lead to an unwieldy situation because of the numbers involved. In such cases, it is recommended that a series of sessions be held instead, with each session being attended by supervisors from like operations. For example, separate meetings could be held for supervisors involved in marketing, design, materials control (including procurement), feeder operations, assembly operations, etc.

As with audit clients and customers, a questionnaire should be provided ahead of time to each auditee supervisor attending such a meeting. This should provide the basic agenda for the discussion. These meetings can be conducted according to the quality circle techniques discussed in Section 15.5.3. A typical questionnaire for this kind of meeting is shown in Figure 15.5.

The data gathered should be analyzed using the techniques outlined in Chapter 12 to determine the trends in audit effectiveness. As all estimates involve confidence levels, it is important that these be clearly stated. They must remain constant if the audit trends are to be charted.

A.B.C. Manufacturing Co. Ltd.
Auditee Evaluation of the Internal Quality Audit

(Please rate each question from 1 to 5 with 5 implying excellent satisfaction.)

Auditee_____

1.0 Were the pre-Audit arrangements carried out to your

satisfaction?

    □   _   □   _   □   _   □   _   □

    1   2   3   4   5

What areas require improvement _____

_____

2.0 Were you fully aware of the objectives and approach to be

taken by the Quality Audit?

    □   _   □   _   □   _   □   _   □

    1   2   3   4   5

What areas require improvement _____

_____

Figure 15.5 Typical auditee questionnaire for use with internal quality audits.

15.4.5 Audit surveillance

Audit surveillance as discussed in Section 15.3.3 can be useful in gathering data on the effectiveness of internal quality audits.

15.4.6 Separate audits

Independent audits conducted by the audit manager can be used to compare his or her findings with those determined by the regular quality auditor(s). It must be recognized that this technique may produce findings somewhat different from those determined by the regular auditor(s) due to the risks involved in sampling procedures. Comparison of the two sets of findings will give an indication of the effectiveness of the regular auditor(s). However, in spite of this drawback, I believe it can be useful in evaluating critical activities or those involving a subjective judgment of conformance.

15.4.7 Auditee performance

Often the results of an audit can be compared with data derived from various verification activities. Thus the verification activities can provide evidence on the effectiveness of the techniques being used to audit the quality system. Where a review of verification problems shows shortcomings in the quality system, the audit results must be examined to see if they have detected the same shortcomings. This review may identify areas in the audit corrective action cycle that require improvement. The effectiveness of any improvements can be verified by future verification data.

15.4.8 Auditor evaluations

As discussed in regard to external audits, so with internal audits postmortem meetings held with the auditors themselves to review their techniques can be useful tools for evaluating audit effectiveness. But with internal quality audits, these meetings should be held periodically throughout the year. As an internal audit is frequently a continuous process, it will seldom be practical to hold this meeting at the completion of a particular audit.

The audit manager will seldom be the font of all audit knowledge; rather different members of the audit team will have different areas of expertise and views on the various elements of the audit process. One of the challenges of being the audit manager is successfully bringing out this knowledge, so every auditor can benefit from it.

For internal audits, it is strongly recommended that quality circles or other participatory techniques be used to determine and evaluate problems and develop solutions. This gives the auditors an added sense of involvement and accountability for the success of the audits. Participatory techniques make it possible to take advantage of all the skills and experiences within the audit group. Quality circle techniques as applied to auditors are discussed in Section 15.5.3.

15.5 Improving the Audit Process

15.5.1 Introduction

The improvement of any process is a creative activity involving free creative thinking by all the individuals involved. In this case, the audit manager serves as the facilitator or impressario who must encourage members of the team to participate to the best of their abilities, and even entice them to exceed their normal expectations. Out of this creative thinking may come a plethora of ideas that will excite further creative thinking by the participants. The activities involved differ from traditional quality circle methods in that the investigation is directed toward solving specific problems; the approach is not wide open. However, the freedom of choice is still very broad across the range of activities and interactions involved with improving quality audit effectiveness.

The quality circle methods discussed in Section 15.5.3 can also serve as growth or training media for both auditors and auditees.

15.5.2 Training

In addition to the basic principles and techniques of auditing, quality auditors should be trained in the following:

1. Differing philosophies on quality assurance, quality control, and process control systems.
2. Different international, national, and industrial procurement quality system standards and guides.
3. A selected range of problem-solving techniques.
4. A thorough grounding in the statistical techniques used in sampling plans, control, and analysis. This should include a knowledge of how to use and abuse statistical data. The training in the abuse of results and techniques is to assist the auditors in detecting incidents of abuse.

Training in these fields is offered by most of the professional and technical societies associated with the auditing, quality, and reliability professions. Courses are also offered by some colleges and universities, and individually by consultants in the field. However, in all cases, the course content should be closely examined to ensure the approaches taken support a constructive mode of operation rather than a negative or punitive approach. Some audit courses imply that the purpose of quality auditing is to identify wrongdoers.

I would also strongly recommend that auditors, auditors-in-training, and managers of audit activities read very broadly, both in the literature related to their own particular discipline and in that having to do with other quality disciplines. In this way, they will become aware of different approaches to the topic of quality. The bibliography at the end of this book gives some indication of the variety of literature available. It indicates the sources found useful by the author in the development of quality assurance systems, and in planning and carrying out quality audits in particular.

15.5.3 Quality circles

15.5.3.1 Introduction. The quality circle is a problem-solving technique that utilizes employee participation to improve the performance of an activity. It provides an excellent means of determining and solving the problems that frustrate employees or prevent them from doing the job right the first time. A successful quality circle program should improve employee morale and efficiency, which in turn should result in an improved product or service and improved productivity. However, these latter effects are the result of an effective quality circle program, and not its objective.

Quality circles per se are no simple panacea to the general problems of an industry. Their success depends, to a large degree, on the management philosophy and working environment of the organization concerned. Unless the management of the organization has developed a climate of employee participation from the top down, a quality circle or other participatory program will solve no problems.

There is a vast array of literature on quality circles and other forms of participatory programs. Therefore, I will not cover the techniques used in these programs in detail, but rather provide some ideas as to how they can be applied to the operational side of conducting a quality audit and the development of improved audit techniques.

15.5.3.2 Operational applications. As indicated in Sections 15.3 and 15.4, it is essential that those conducting an audit be aware of any problems involving its actual operations. These may be problems that arise during the course of the audit or afterward, in relation to the reports issued by the auditors. Here we are speaking about problems associated with the operation of the audit, *not* any quality system problems uncovered by the audit. The individual questionnaires shown in Figures 15.2 through 15.5 serve to collect basic information about the audit from the questionnaire respondents. However, the answers given by those individuals may not reflect the view of the majority on the receiving end of the audit. For this reason, it can be of great assistance to bring together a number of representatives of the client(s) or auditee(s) to participate in a session aimed at problem identification.

In scheduling such a session, it is best to have the audit function represented by its manager, who will act as the facilitator or catalyst to bring out ideas for improvement, rather than the individual auditors. Providing the manager demonstrates that the objective of the session is to improve the efficiency of the audit operation, a free interchange of thoughts should occur. The presence of individual auditors could inhibit the development of this open climate. The audit manager must, of course, be well briefed on any problems encountered by the audit staff.

Sessions of this nature may be difficult to arrange in relation to external quality audits, except in cases where the client is the procurement or purchasing department of an organization and the auditees are the potential suppliers being evaluated. Then a meeting involving buyers and those responsible for the verification of purchased goods and services could be of great value. If a good vendor/vendee relationship exists, it may be possible to bring a number of suppliers together in a single session.

With internal quality audits, such sessions can readily be developed. Each meeting should involve a particular level of management and/or employees involved in related activities. For example, the chief executive officer and senior managers would not normally participate in the same session as middle and lower management personnel, since the needs and expectations of both these groups will undoubtedly be different. Sessions involving middle and lower levels management would normally be grouped by related activities. In a large organization, separate sessions may be set up by department—marketing, design, materials, control, production, etc. In a smaller organization, all the first-line supervisors may attend a single meeting regardless of their areas of responsibility.

15.5.3.3 Improving audit techniques.
Participatory sessions on improving quality audit techniques will involve the individual auditors, with the audit manager serving as the facilitator or catalyst necessary to bring forth creative ideas. Individuals from outside the audit function should not normally be involved in these sessions. This does not imply that an improved technique should be introduced without reference to those to whom the audit is to be applied. Instead, solutions to problems should be reviewed with both client(s) and auditee(s) prior to their implementation.

The agenda for a session of this nature would consist of three major segments:

1. Identification of operational problems noted by or bothering the auditors.
2. Development of a composite list of the improvements required, using the list resulting from step 1 above and those developed during the operational sessions.
3. Development and verification of solutions to the problems on the composite list.

In developing the composite list, it must be recognized that many of the problems identified may actually be symptoms of a more fundamental cause or problem. Therefore, it is essential that the composite list always indicate what are the true problems and what are their related symptoms. Solutions can then be developed for each problem, using the combined skills of the audit team. As with other solutions to process problems, the proposed corrections must be verified by some means before they are finalized. Part of this verification can be a review of the revised techniques in relation to activities giving rise to the original problems. This will help participants recognize how the audit is a positive reinforcement process used to prevent problems from becoming catastrophic and to identify any hidden adverse interactions.

15.6 Conclusions

The philosophy of improvement is an essential element in the quality audit process, as it is in all other elements of an organization. In effect, any technique is obsolete once it has been implemented and is apparently working smoothly, since it is then ready to be analyzed in terms of improvement. And once a process or technique is ready for improvement, it is ready for an improvement program.

In putting these thoughts together, I have attempted to provide a range of ideas, techniques, and approaches that will stir the imaginations of quality auditors, clients, and auditees alike. Out of this will come new and more constructive objectives for quality audits and new and better ways of conducting them.

16

Conclusions

16.1 An Audit Truism

All audits should provide a positive reinforcement and learning experience to all participants, i.e., to clients, auditors, and auditees alike. The auditees and auditors experience the most opportunities for these benefits.

Positive reinforcement occurs for an auditee when strong points in the quality system are recognized by the auditor and commented on during the audit and in the reports. It occurs for an auditor when the findings enable corrective action being taken to prevent troubles or problems from arising. What stronger form of positive reinforcement can be found than helping keep someone out of trouble?

The learning experience results from the opportunity of observing first-hand different interpretations of requirements and the methods that must be used to meet the requirements. This applies to all three parties, although again the greatest scope is available to auditees and auditors.

The audit should be welcomed by all parties as an opportunity to gain knowledge rather than seen as an imposition aimed at entrapping the auditee on some minor point.

16.2 Introduction

As we approach the final decade of the twentieth century, quality audits provide one of the most interesting challenges to quality practitioners. One aspect of that challenge is the need such an audit imposes to consistently judge the merits of each work situation against the relevant reference standard and yet remain broad- or open-minded about the mechanics of meeting the standard's requirements. This involves the ability to recognize the intent of the standard and reconcile it with requirements of each work situation. The other challenging aspect of quality audits comes from the recognition that a key objective for most of them is to help the auditee reduce the risks inherent in quality or reliability problems—it is the auditor's job, in other words, to make a quality audit a constructive process and not a destructive one.

16.3 People Orientation

A quality audit is a people-oriented activity since it evaluates the acts and/or decisions of individuals with respect to a reference standard and/or the documentation of a quality system. Ascertaining product or service conformance, on the other hand, is a product- or function-oriented verification activity, i.e., an inspector, by whatever title— drawing checker, tester, proofreader, test engineer, product or service evaluator, etc.—is used to verify a particular product or work function.

Properly applied, a quality audit is the means by which the quality organization, by whatever name it is known, provides positive reinforcement to the line personnel and those engaged in the support activities to those main work functions. It also is the means of determining the effectiveness of the quality system of an organization with respect to some external standard, whether the standard be international, national, industrial, or other.

16.4 Quality System Improvement and the Quality Audit

Improvement of operating systems is fast becoming one of the prime objectives of successful quality systems. An effective improvement program results in more efficient production of a higher-quality product or service. In this way, a good quality system improves the status of the organization in the eyes of its customers, while improving the cost and delivery performance against internal standards. The improvement program involves all the organization's personnel, from the chief executive officer down.

The quality organization should be one of the prime instigators in implementing an improvement program. Therefore, it should set an example in the application of the creative process to improving the activities involved in the quality system. A key operation in a quality organization is carrying out quality audits. As a highly visible tool used to evaluate system effectiveness, it must demonstrate its usefulness by applying dynamic, creative thinking to its own process through improvements in its own activity and identifying potential improvements in other areas.

16.5   Organization Relationships

16.5.1   Audit/Client relationship

A quality audit provides the client with an independent assessment of the conformance and effectiveness of an organization's operating system against predefined standards. As discussed in Chapter 3, the relationships between the client, the auditors, and the auditee can be varied.

16.5.2   Auditor/Auditee relationship

The relationship between auditor and the auditee determines one aspect of the audit—whether it is internal or external in nature. An internal quality audit evaluates conformance to an organization's operating procedures and the effectiveness of these procedures. An external audit can do this also, but in addition determines conformance to some predefined standard.

16.5.3   Auditee personnel/Audit relationship

Although the reference standards involved to a large extent also determine the type of audit to be carried out—quality system, management, product, service, software, process, etc.—quality audits are basically concerned with the acts or decisions of the people doing the activities being audited. The output of the activities, whether product, service, hardware, or software, will be evaluated against the reference standard's requirements. However, the evaluation is not aimed at determining the acceptability of the output itself, but rather at determining how the output is achieved so that the individual decisions that ultimately will result in the output can be evaluated. Thus the evaluation of the output itself is merely a means to an end, not an end in itself. Determining the acceptability of the output per se is an inspection function, regardless of the title or skills of the individuals making that determination.

16.6   The Auditor

16.6.1   Auditor status

In all auditing situations, the auditors act on behalf of a client who is frequently the executive officer of the organization concerned but always a member of the senior management team. As an audit progresses, it evaluates how well the various managers are fulfilling their responsibilities and accountabilities. Hence auditors should have a level of skills and knowledge compatible with the level expected of the most senior management positions being audited. In addition, they need a thorough knowledge of the intent and application of all aspects of the quality system being audited, and the skills of a quality engineer. Therefore, quality auditors should be qualified to be part of the management team, and be recognized as having those qualifications. Auditing is *not* a make-work activity for surplus inspectors or other more junior positions.

16.6.2 Auditor knowledge

An auditor must be well versed in all aspects of the quality profession, including its reliability, availability, and maintainability (RAM) aspects. This knowledge is necessary to adequately evaluate the various elements of quality systems. Without such thorough knowledge, an erroneous evaluation may be made of an unusual application of a technique. It is recommended that auditors become active in their national society for quality practitioners as well as international societies in the field. This is particularly important with respect to those societies involved in the preparation of national and international standards in the quality and reliability fields.

Auditors should also be knowledgable in such peripheral fields as accounting principles, industrial engineering principles, and applications of computer techniques. They will find these of value in evaluating the effectiveness of the various processes they encounter in their work, as well as an aid to implementing improvement programs in regard to the audit function itself.

It is also essential that auditors have a broad knowledge of the industrial or service discipline(s) with which they will be concerned. They need not be specialists in particular fields, but rather generalists with some knowledge of all the aspects involved in the discipline(s). Whenever possible, auditors should qualify for professional standing in such disciplines, either by acquiring an appropriate degree or by obtaining a recognized level of membership in the appropriate professional societies.

16.7 Planning a Quality Audit

As a member of the management team, an auditor must have a thorough knowledge of the planning and system analysis tools available for and applicable to the area of concern. Typical approaches and tools were discussed in Chapters 4 through 6 and 8 through 10. An auditor needs a battery of tools in order to be able to optimize the choice for a particular application. His or her ability to plan must be thorough in order to reduce the possibility of friction arising during an actual audit. An auditor should always work from some form of working papers (Chapter 9) to ensure that all the necessary points are covered by the audit and no unnecessary duplication of effort occurs.

16.8   Implementation of a Quality Audit

The implementation of a quality audit is an exercise in personal relationships that involve clients of various types as well as all the levels of the personnel hierarchy of the auditee. The upper echelon personnel of the auditee are involved in the initiation, scheduling, and initial briefing meetings before the audit commences (Chapter 5) and the debriefing meeting at its conclusion (Chapters 11 and 14). Middle and lower management levels are involved during the course of the audit, as are the professionals, technologists, operators, etc., who actually carry out the activities being audited. Thus an auditor must be prepared to deal with personnel having a wide variety of knowledge and skill levels.

The actual audit presents a dichotomy between the persistent search for facts and the diplomacy and tact necessary in dealing with people. An auditor needs to be that strange mixture one might term a "stubborn diplomat." However, the stubbornness must not lead to a closed- or narrow-minded approach to the mechanics of meeting system requirements.

An auditor must be prepared to use all the senses—hearing, smell, taste, touch, and sight. The prime sensors will be the ears and eyes as they monitor the results of the questions posed to the auditee. However, care must be taken that the eyes do not suffer from tunnel vision or the ears from narrow-band hearing. Peripheral sensing is important in both hearing and seeing. Much can be learned from an awareness of surrounding activities as well as from attention to the topic under review. The situation is very much akin to a driver keeping his or her visual attention acute by taking note of what is going on alongside or over the road. And similarly the other senses should be noting evidence as much from peripheral happenings as from the point under review. Peripheral sensing frequently gives the first indication that something is not quite right in the general area of concern.

16.9   Analysis of Audit Findings

Problem solving and statistical analysis are the two keys to analyzing the findings of a quality audit.

Problem-solving techniques are used to determine the actual cause behind the problems observed during an audit. These problems are frequently merely symptoms, and their underlying cause must be identified and corrected if the problems are not to recur. Auditors should be able to use several different problem-solving techniques so the optimum one can be chosen for a particular application. Auditors should also welcome the opportunity to evaluate and use new techniques suggested by auditees.

Statistical techniques are used in deriving data during the audit and analyzing that data after the audit. They are frequently needed because the number of potential observations in most areas being audited is far too great for each one to be checked directly. Therefore, overall performance is estimated based on the findings within a particular sample.

It should be noted, however, that all sampling procedures involve the inherent risk of failing to detect at least one example of the various types of errors present. But a judicious selection of the sampling plan to be used can match this risk to the criticality of the topic being audited.

Statistical tools are used in the analysis to quantify the overall performance of particular areas of the auditee's facility based on the data derived during the audit. These calculations always involve confidence levels. The confidence level must always be stated when giving a quantified performance figure.

16.10   Reporting a Quality Audit

A quality audit report, in whatever form it is issued, is the raison d'être of the audit itself. The sole purpose of holding an audit is to provide information or data to the auditee and/or client. Different auditees and clients have different outlooks, abilities, interests, and skills, and auditors must be able to prepare effective verbal or written reports that convey the necessary information to the recipient(s).

Reports must be designed to meet the needs of the customer, i.e., the client. With the client's permission, reports may also be issued to the auditee. Internal audit reports cover both the effectiveness and conformance of the quality system being audited, as well as the necessary corrective actions required to improve the system or correct identified shortcomings.

Reports interpret the findings of an audit. They are *not* listings of the audit's observations. In all cases, reports must distinguish the observations of problems from their underlying cause(s).

In order to produce effective reports, it is recommended that auditors have word-processing facilities available and be well versed in the capabilities of the system being used. Much time can be saved by having entry and printing facilities available in the audit office. With the abilities of current and anticipated future generations of personal computers, the audit office can easily be self-contained.

If auditors do not have word-processing facilities available for their use, they should be strongly encouraged to develop their abilities to dictate letters, reports, etc., for transcription elsewhere. This practice provides two benefits over handwritten originals, namely:

1. The writer is forced to plan the document before actually starting its detailed composition. Such planning normally takes the form of listing the points to be covered in the order in which they are to be addressed.

2. Dictation normally takes far less time than writing out letters or reports by hand, even allowing for the more careful planning involved.

16.11 Corrective Action

Motivating corrective action is a major objective of all quality audits. Despite the wording of some quality standards, the audit team cannot be responsible for taking corrective action. That lies with the individual or group responsible for the activity, product, or service needing correction.

Motivating corrective action can be a very sensitive area since many individuals become very protective of existing techniques when they are criticized from outside their own group. This becomes particularly serious when the activity involves original work by the individual concerned, e.g., original design work done by a designer, a manufacturing method initiated by a particular industrial engineer, etc.

Requests for corrective action are discussed in Chapter 13. Request formats can vary; however, a request should always be documented and a written reply required.

16.12 Quality Audit Management

Quality audit management must have some means of determining how well auditors are carrying out their activities. In effect, there must be a measurement or audit system in position to audit the performance of auditors and analyze the finding. A number of techniques for doing this are discussed in Chapter 15.

One cautionary note—the audit manager should *never* make a decision on behalf of an indecisive auditor. The decision must come from the individual auditor, who may request a review of that decision. If this type of incident occurs, it is recommended that the manager require the auditor to justify the decision. The exercise of justifying the decision will help the auditor clarify some of his or her decision-making problems.

An audit manager must encourage creativity among auditors in regard to developing new and improved audit techniques.

16.13   Quality Audit/Reference Standards

Reference standards and guides affecting a quality audit fall into two basic categories:

1. Those used as the benchmarks for measuring the effectiveness of the quality system being audited
2. Those used to define how quality audits should be performed and by whom

Quality system standards in the first category are issued by many of the national standards writing organizations (SWOs) around the world, as well as certain of the international SWOs. Appendix A in Chapter 1 provides audit-related definitions or requirements as defined in selected English language standards in this field. The national standards reviewed there are issued by the United States, Canada, the United Kingdom, and Australia. The international ones are issued by the International Organization for Standardization and the North Atlantic Treaty Organization. The details of source for these are given in Appendix 1A.7. Variations of these standards are used by many other countries. With the growth of international trade and multinational organizations, it is rapidly becoming necessary for auditors to be thoroughly familiar with the standards issued by countries other than their own, and in particular those issued by international organizations such as ISO and NATO.

Regarding the second category, currently the only English language standards on quality auditing are issued by the United States and Canada, (see Chapter 1, Appendix 1A.7, for details). ISO is developing a series of audit standards using these documents as their foundations. The ISO series should be ready in the near future and will undoubtedly be adopted by many of the countries participating in the ISO Technical Committee TC 176 (see Chapter 1, Appendix 1B, for the current membership list).

Again, auditors should be familiar with their own nation's, other nations', and international quality auditing standards.

16.14   Measurement of Auditor Performance

Techniques for measuring auditor performance are discussed in Chapter 15. They are discussed there in terms of determining the effectiveness of the audit activity. However, as standards and/or certification programs become available defining minimum requirements for auditors with respect to formal education, professional knowledge, independence from auditee influence, etc., new techniques will have to be developed for measuring the conformance of the auditors to these requirements. Third-party certification of auditors can be an effective way of doing this, and the Quality Auditor Certification program of the American Society for Quality Control is the first of these granting general certification in operation. Acceptance of ASQC's certification will grow as it develops credibility. Based on previous ASQC certification programs, this should happen rapidly. Agencies such as the British Standards Institute and the Canadian Standards Association have certification programs to qualify those auditors evaluating quality systems on their behalf.

16.15   Auditor Corrective Action

A number of different techniques are available for improving auditor performance. These include:

1. Training courses on auditing

2. Conferences, seminars, etc., on quality systems and auditing

3. Quality circle programs functioning within an audit or other organization

4. Discussion groups on the topic within a given locality or professional or trade organization

5. Participation in standards writing committees

The value of the first two of the above are self-evident. However, their practicality will depend on the location in which they are offered and to some extent on the discipline with which the individual auditor is involved.

Quality circle activities lend themselves to the improvement of audit activities. By pooling ideas on the problems encountered in auditing, participants can call on the full strength of the audit organization to find efficient solutions. Normally these circles consist of the auditors and their supervisors. However, when problems encountered in internal quality audits are the area of concern, there can be value in bringing in representatives of the auditee. This can allow problems to come to light from both sides of the audit situation. Chapter 15 discusses these approaches in general. There is also considerable literature available on quality circle techniques.

Discussion groups (item 3 above) are similar to quality circles in that the experience of participants is pooled to broaden the knowledge of everyone. As with quality circles, solutions can be developed using the full resources of the group. Since discussion groups involve auditors from different organizations, they tend to provide a broader variety of knowledge than the QC. However, the technique is somewhat restricted by the need to protect the confidentiality of client and auditee data. Discussion groups can be developed by local chapters or sections of professional or trade societies as a means of improving the performance of auditors.

Participation in the development of national and international quality system and quality audit standards (item 5 above) also exposes auditors to a broad variety of different approaches. This helps broaden the outlook of auditors as well as provide them with background knowledge on the intent of quality system standards. However, care must be taken that this participatory approach to standards does not inhibit the broad-minded approach so necessary to be a successful quality auditor.

16.16 Conclusions

I have attempted to provide ideas and thoughts that will enable the readers of this book to take advantage of my audit experience, which includes over 20 years of involvement with managing and implementing both external and internal quality audits. My intention has been to motivate readers to find better and more useful ways of approaching their own audit work. Thus I have tried to provide a foundation on which readers can continue to build. This aim also illustrates one of the reasons I was active in the formation of the Quality Audit Technical Committee of the American Society for Quality Control: I saw the opportunity that it gave to help in the development of individuals working in and techniques applicable to this important field.

Quality Audit Bibliography

Introduction

This bibliography lists some of the texts, articles, and standards that I and my colleagues have found useful over the years in developing, applying, and improving our audit techniques. It is not intended to be exhaustive.

Part 1. Texts

1. Juran, J. M., and F. M. Gryna, *Quality Control Handbook,* 4th ed. New York: McGraw-Hill, 1988.
2. Juran, J. M., and F. M. Gryna, *Quality Planning and Analysis,* 2nd ed. New York: McGraw-Hill, 1980.
3. Feigenbaum, A. V., *Total Quality Control,* 3rd ed. New York: McGraw-Hill, 1983.
4. Sinha, M. N., and W. Willborn, *Management of Quality Assurance.* New York: John Wiley & Sons, 1980.
5. Stettler, H. F., *Systems Based Independent Audits,* 5th ed. Englewood Cliffs, NJ: Prentice-Hall, 1982.
6. Sayle, A. J., *Management Audits.* Milwaukee: Quality Press, 1985.
7. Janis, Irving L., and Leon Mann, *Decision Making.* New York: Free Press, 1979.
8. Willborn, W., *Compendium of Audit Standards.* Milwaukee: Quality Press, 1983.
9. Johnson, L. M., *Quality Assurance Program Evaluation.* Whittier, CA: Stockton Doty Press, 1970.
10. *Workbook* for item 9.
11. Harris, D. H. and F. B. Chaney, *Human Factors in Quality Assurance.* New York: John Wiley & Sons, 1968.
12. Publications of the Institute of Internal Auditors, Inc., 232 Maitland Ave., Aliamonte Springs, FL 32701.:
    a. Systems Auditability and Control, 1977
    Part 1. Executive Report
    Part 2. Control Practices
    Part 3. Audit Practices
    b. Sampling Manual for Auditors, 1st ed., 1967
    c. Supplement to the Sampling Manual for Auditors, 1970
    d. Behavioral Patterns in Internal Audit Relationships, 1972
13. Publications of the Canadian Institute of Chartered Accountants, 250 Bloor Street, Toronto, Ontario, Canada M4W 1G5:
    a. Materiability in Auditing, 1974

b. *Independence of Auditors,* 1976
c. *Statistical Sampling in an Audit Context,* 1975
d. *Good Audit Working Papers,* 1970
e. *The First Audit Engagement,* 1975
f. *Internal Control and Procedural Audit Tests,*1977

14. Arkin, H., *Handbook of Sampling for Auditing and Accounting.* New York: McGraw-Hill, 1984.
15. Skinner, R. M., and R. J. Anderson, *Analytical Auditing.* New York: Sir Isaac Pitman, 1966.
16. Publications of the American Institute of Certified Public Accountants, Inc., 1211 Avenue of the Americas, New York, NY, 10036:

   a. *The Auditor's Reporting Obligation,* 1972
   b. *The Auditor's Study and Evaluation of Internal Control of EDP Systems,* 19770

Part 2.   Articles

2.1   Publications of the American Society for Quality Control, 310 West Wisconsin Avenue, Milwaukee, WI 53203-9990

2.1.1   Analysis

1. Stephens, Dick, and Dave Fox, "IDEF Modeling Application to Quality Assurance Assessment," *1987 ASQC Quality Congress Transactions,* pp. 512–517.
2. Bailie, Howard H., "Quality Performance Measurement," *1985 ASQC Quality Congress Transactions,* pp. 15–20.
3. Messina, William S., "A Tool for Measuring Manufacturing Quality," *1983 ASQC Quality Congress Transactions,* pp. 40–44.
4. Puri, Subhash C., "Quality Indicators for Corporate Management," *1983 ASQC Quality Congress Transactions,* pp. 635–640.
5. Hsiang, Thomas C., and John J. Gordon, "New Statistical Methodologies in a QA Audit System," *1982 ASQC Quality Congress Transactions,* pp. 335–342.
6. Mills, Charles A., "Risk and Human Resources are Factors in Sampling," *1982 ASQC Quality Congress Transactions,* pp. 619–625.
7. Lee, Martin L., Julia Kantrowitz, and Mary Taylor, "Statistical Validation of Quality Control Tests," *1982 ASQC Quality Congress Transactions,* pp. 717–724.
8. Mills, Charles A., "The Risk Factor in the Quality Audit," *1980 ASQC Technical Conference Transactions,* pp. 454–459.
9. Coswell, Alan R., and J. D. Zimmer, "Flow Chart Analysis of Pharmaceutical Products," *1979 ASQC Technical Conference Transactions,* pp. 161–167.
10. Hansel, John L., "Corrective Action and Pareto—The Perfect Marriage," *1979 ASQC Technical Conference Transactions,* pp. 185–189.
11. Puri, Subhash, and John R. McWinnie, "Quality Management through Quality Indicators: A New Approach," *1979 ASQC Technical Conference Transactions,* pp. 710–713.
12. Slack, Thurman J., "Quality Performance Can Be Rated," *1978 ASQC Technical Conference Transactions,* pp. 347–352.
13. Burns, Robert G., "Trend Analysis—The Key to Corrective Action," *1978 ASQC Technical Conference Transactions,* pp. 386–389.
14. Shainin, Dorian, "Creative Industrial Quality Problem Solving," *1977 ASQC Technical Conference Transactions,* pp. 83–84.
15. Hill, R. M., "Problem Solving Techniques," *1977 ASQC Technical Conference Transactions,* pp. 519–529.
16. Kenney, James M., "Hypothesis Testing: Guilty or Innocent," *Quality Progress,* January 1988, pp. 55–57.

2.1.2 Application

1. Kohnen, James B., "Food Processing System Audit Techniques," *1987 ASQC Quality Congress Transactions*, pp. 65–68.
2. Whittingham, P. R. B., "Operator Self Inspection," *1987 ASQC Quality Congress Transactions*, pp. 278–286.
3. Marroquin, Pedro S., "Quality Control System Analysis," *1987 ASQC Quality Congress Transactions*, pp. 359–369.
4. Davis, Stephen L., "Conversion to Quality Audit System," *1985 ASQC Quality Congress Transactions*, pp. 25–27.
5. Cattaneo, Donald J., "Health Care Industry Validation in the 1990's," *1985 ASQC Quality Congress Transactions*, pp. 325–333.
6. Marash, I. Robert, "The Psychology of Performing a GMP Audit," *1985 ASQC Quality Congress Transactions*, pp. 336–341.
7. Kane, Roger W., "Fitness Reviews: Key to a Total Quality Program," *1984 ASQC Quality Congress Transactions*, pp. 263–267.
8. Goldstein, Raymond, "The Two-Tier Audit System," *1983 ASQC Quality Congress Transactions*, pp. 14–16.
9. Chipman, Laurence D., P. E., "Quality System Auditing to Meet Medical Devices GMP's," *1983 ASQC Quality Congress Transactions*, pp. 557–559.
10. Love, Kenneth S., "Quality Auditing of New Products," *1982 ASQC Quality Congress Transactions*, pp. 534–537.
11. Miller, Ervin F., "Corporate Quality Audit/Survey," *1982 ASQC Quality Congress Transactions*, pp. 538–545.
12. Thresh, James L., "New Audit Trends—Key to Productivity and Profits," *1982 ASQC Quality Congress Transactions*, pp. 546–551.
13. Davis, S. R., "Energy Audits and How to Reduce Operating Costs," *1980 ASQC Technical Conference Transactions*, pp. 193–200.
14. Besterfield, Dale H., "A Model for a Product Safety Process Audit," *1979 ASQC Technical Conference Transactions*, pp. 499–505.
15. Diggs, H. J., "Why Have Outgoing Quality Audits," *1979 ASQC Technical Conference Transactions*, pp. 506–512.
16. Cogue, Jean-Marie, "Quality Audits in French Industry," *1979 ASQC Technical Conference Transactions*, pp. 594–597.
17. Dempsey, William, "Auditing the Company that 'Tries Harder,' " *1979 ASQC Technical Conference Transactions*, pp. 598–604.
18. Schock, Harvey E., Jr., "Regulatory Compliance and Audit/Survey," *1979 ASQC Technical Conference Transactions*, pp. 605–607.
19. Willborn, Walter O., "Quality Audits in Support of Small Business," *1978 ASQC Technical Conference Transactions*, pp. 179–185.
20. Mason, Leon V., "Is Your QA Organization Aired at the Right Time?," *1978 ASQC Technical Conference Transactions*, pp. 264–267.
21. Levine, Ralph I., "Quality System Audit," *1978 ASQC Technical Conference Transactions*, pp. 401–408.
22. Eyres, Ron, "Product Audit for High Volume Low Cost Inspection," *1977 ASQC Technical Conference Transactions*, pp. 53–56.
23. Zeccardi, Joseph J., "Auditing Systems which Affect Product Quality," *1976 ASQC Technical Conference Transactions*, pp. 323–329.
24. Alainio, A. F., "Quality System Audit in a Multi-Plant Manufacturing Division," *1972 ASQC Technical Conference Transactions*, pp. 211–218.
25. Palmer, D. J., "The Real Pay-offs of Quality Audit," *Quality Progress*, June 1977, pp. 28–31.
26. Sheppard, John W., "CASE Is Now 12 Years Old," *Quality Progress*, November 1977, pp. 30–34.
27. Golomski, W. A., "Operational Auditing," *Quality Progress*, September 1968, pp. 34–35.
28. "Auditors Foster Compliance," *Quality Progress*, October 1978, pp. 10–11.

29. Binstock, S. L., "Certification Program Aims at Quality Standards in Steel Construction," *Quality Progress*, November 1978, pp. 18–20.
30. Dine, H. A., "Quality Auditing—A Familiar Land Revisited," *Quality Progress*, November 1978, pp. 34–37.
31. Mills, Charles A., "In-Plant Quality Audit," *Quality Progress*, December 1976, pp. 22–25.
32. Willborn, Walter, "A Generic QA Audit Guideline," *Quality Progress*, January 1987, pp. 24–25.
33. Pereira, Armando Lopes, "Quality Audits and International Standards," *Quality Progress*, January 1987, pp. 27–29.

2.1.3 Human resources

1. McConnell, Nancy, "Reinforcing the Positive Paves the Way for Success," *Quality Progress*, January 1987, p. 12.
2. Hayton, Thomas, "The Problem of Measuring Human Performance," *1987 ASQC Quality Congress Transactions*, pp. 734–739.
3. White, Quitman, Jr., "Audits Contribute to Pride, Productivity and Profit," *1984 ASQC Quality Congress Transactions*, pp. 266–272.
4. Jaehn, Alfred R., "Measuring Employee Quality Performance," *1982 ASQC Quality Congress Transactions*, pp. 914–917.
5. Thomas, E. F., "People: Cause and Effects," *1976 ASQC Technical Conference Transactions*, pp. 185–195.
6. Vassalo, J. R., "Psychology of Auditing," *Quality Progress*, September 1977, p. 37.
7. Hershauer, James C., "An Audit of Employee Attitudes about Quality," *Quality Progress*, August 1979.

2.1.4 Implementation

1. Lay, Henry G., "The Quality Audit: Who Does What," *Quality Progress*, January 1987, pp. 20–22.
2. Adams, Ray, "Moving from Inspection to Audit," *Quality Progress*, January 1987, pp. 30–31.
3. Plum, Kathryn S., "A Success Story," *Quality Progress*, January 1987, pp. 32–34.
4. Willborn, Walter, "Software Quality Auditing," *Quality Progress*, January 1987, pp. 36–37.
5. Farrow, John H., "Quality Audits—Control Coefficient in the Equation," *1987 ASQC Quality Congress Transactions*, pp. 507–511.
6. Farrow, John H., "Quality Audits—Road to Teamwork," *1985 ASQC Quality Congress Transactions*, pp. 21–24.
7. Parier, Anil S., "Quality Audit Implementation," *1983 ASQC Quality Congress Transactions*, pp. 188–190.
8. Hehl, Mark C., "Assurance Compliance through a Three Tier Audit Program," *1983 ASQC Quality Congress Transactions*, pp. 191–196.
9. Greenwell, Warren D., "Squeezing the Most from Your Audit Dollar," *1981 ASQC Quality Congress Transactions*, pp. 1–7.
10. Scott, Case A., "Results Oriented Auditing," *1981 ASQC Quality Congress Transactions*, pp. 8–10.
11. Sternberg, Alexander, "Quality Assurance Auditing—What It Should Be," *1980 ASQC Technical Conference Transactions*, pp. 446–453.
12. Willborn, Walter O., "Quality Audits: Standards and Concepts," *1980 ASQC Technical Conference Transactions*, pp. 522–526.
13. Hill, William J., "Improving Audit Effectiveness," *1978 ASQC Technical Conference Transactions*, pp. 259–263.
14. Marash, Stanley A., "Performing Quality Audits," *Industrial Quality Control*, 1962, pp. 342–347.
15. Marguglio, B. W., "Quality Systems Audit," *Industrial Quality Control*, 1963, pp. 12–15.

2.1.5 Management-oriented

1. Farrow, John H., "Quality Audits: An Invitation to Management," *Quality Progress*, January 1987, pp. 11–13.
2. Burr, John T., "Overcoming Resistance to Audits," *Quality Progress*, January 1987, pp. 15–18.
3. Ishikawa, Kaoru, "The Quality Control Audit," *Quality Progress*, January 1987, pp. 39–41.
4. Shimoyamada, Kaoru, "The President's Audit: QC Audits at Komatsu," *Quality Progress*, January 1987, pp. 44–49.
5. Marash, I. Robert, "Help I'm Being Audited by the FDA," *1987 ASQC Quality Congress Transactions*, pp. 169–173.
6. Dorsky, Lawrence R., "Validation—A Do-it-Yourself Top Management Kit," *1987 ASQC Quality Congress Transactions*, pp. 833–839.
7. Birmingham, Fletcher A., "Audit to Standards for Excellence in Quality," *1986 ASQC Quality Congress Transactions*, pp. 187–192.
8. Callner, Phillip D., "Quality Excellence—Maintenance by Management Audit," *1986 ASQC Quality Congress Transactions*, pp. 193–198.
9. Willborn, Walter O., "Quality Audits and Technological Excellence," *1986 ASQC Quality Congress Transactions*, pp. 199–205.
10. Farrow, John H., "Quality Audits—Why Bother," *1984 ASQC Quality Congress Transactions*, pp. 259–262.
11. Farrow, John H., "Quality Audit—Means to Management Commitment," *1983 ASQC Quality Congress Transactions*, pp. 10–13.
12. Marash, Stanley A., P. E., "Quality Auditing as a Management Function," *1983 ASQC Quality Congress Transactions*, pp. 17–19.
13. Sinha, Madhay N., and Walter O. Willborn, "Total Quality Audit: A New Approach," *1983 ASQC Quality Congress Transactions*, pp. 197–201.
14. Willborn, Walter O., "A Generic Guideline for Quality Audits," *1983 ASQC Quality Congress Transactions*, pp. 327–330.
15. Mills, Charles A., "A Management Challenge to the Quality Professional," *1981 ASQC Quality Transactions*, pp. 175–177.
16. Farrow, John H., "Quality Audits—Now They're Asking for Them," *1980 ASQC Technical Conference Transactions*, pp. 439–445.
17. Reis, P. S., and G. I. Fahrenbruch, "Quality Audit—An Effective Management Tool," *Industrial Quality Control*, February 1966, pp. 402–407.
18. Mills, Charles A., "Interactions of Industrial Integration and Standardization," *Quality Progress*, February 1970, pp. 22–23.
19. Wachniak, R., "Managerial Control—The Quality Audit," *Quality Progress*, November 1975, pp. 22–25.

2.1.6 Planning

1. Freund, Richard A., "QART—A Quality Assurance Status Evaluation," *1983 ASQC Quality Congress Transactions*, pp. 331–336.
2. Duham, Stanley, "An Audit Is More than an Audit," *1979 ASQC Technical Conference Transactions*, pp. 491–498.
3. Sternberg, Alexander, "The Training and Certification of Nuclear Q. A. Auditors," *1978 ASQC Technical Conference Transactions*, pp. 186–195.
4. Ford, K. C., "The Co-Ordination of QA Assessments and Audits," *1978 ASQC Technical Conference Transactions*, pp. 196–199.
5. Mills, Charles A., "Discovery Sampling in an Audit Situation," *1978 ASQC Technical Conference Transactions*, pp. 673–675.
6. Emmons, Sidney L., "Auditing for Profit and Productivity," *1977 ASQC Technical Conference Transactions*, pp. 206–212.
7. Carr, Wendell E., "Inspection Planning," *1974 ASQC Technical Conference Transactions*, pp. 401–406.
8. Law, C. W., "Expanding the Scope of Quality Assurance Audits," *Quality Progress*, October 1970, pp. 31–32.
9. Butler, R. C., "Zero Acceptance Sampling Plans: Expected Cost Increases," *Quality Progress*, January 1988, pp. 43–46.

2.1.7 Supplier-Oriented

1. Johnson, Stanley G., "Theory Y Vendor Evaluation Using Lotus 1-2-3™," *1987 ASQC Quality Congress Transactions,* pp. 702–709.
2. Madigan, Martin J., "Xerox Process Qualifications for Suppliers," *1986 ASQC Quality Congress Transactions,* pp. 96–102.
3. McAdam, Eugene H., "Rating 'Special Process' Suppliers," *1982 ASQC Quality Congress Transactions,* pp. 84–89.
4. Silver, Ben, "How to Perform a Major Sub-Contract Audit," *1980 ASQC Technical Conference Transactions,* p. 825.

2.2 Other Publications

1. Kondo, Yoshio, "Internal QC Audit in Japanese Companies," *EOQC Quality,* Issue No. 4, 1959, pp. 97–102.
2. Ainsworth, L., "The Use of Signal Detection Theory in the Analysis of Industrial Inspection," *Journal of the Institute of Quality Assurance,* September 1980, pp. 63–68. (Address: 10 Grosvenor Gardens, London SW1 W0DQ, United Kingdom.)
3. Gibson, J. D., "How Do You Recognize a Qualified Inspector?" *Journal of the Institute of Quality Assurance,* June 1983, pp. 51 and 56.
4. Stebbing, L. E., "Quality Assurance in the Design of Offshore Structures," *Journal of the Institute of Quality Assurance,* June 1983, pp. 52–56.
5. Gilmore, H. L., "Corporate Concern for the Quality of Employee Performance: Company Practice and Interpersonal Relations," *Journal of the Institute of Quality Assurance,* September 1984, pp. 71–74.
6. Schuuman, F. J. B., "The Selection of Suppliers," *Journal of the Institute of Quality Assurance,* December 1984, pp. 95–98.
7. Monk, J. K., "The Selection and Training of Supplier Assessors," *Journal of the Institute of Quality Assurance,* December 1984, pp. 99–105.
8. Tilley, B. G., "Getting Results with Effective Quality Auditing," *Journal of the Institute of Quality Assurance,* December 1984, pp. 109–111.

Part 3.   Standards

See Chapter 1, Appendix 1A.

Index

About the Author

Charles A. Mills, a member of the American Society for Quality Control, has devoted himself to writing, teaching, and consulting in the field of quality systems since his retirement from Westinghouse Canada, Ltd. He serves as Vice Chairman of the Canadian Standards Association Steering Committee on Managing for Quality and Reliability, and is a member of the Canadian Standards Association Technical Committees on the Quality Audit and Statistical Techniques.